ARITHMETIC &
TRIGONOMETRY

B.P.Brindle, B.Sc.

**Published by Intercontinental Book Productions in
conjunction with Seymour Press Ltd.**
Distributed by Seymour Press Ltd., 334 Brixton Road, London, S.W.9 7AG

1st Edition 3rd Impression © Intercontinental Book Productions 1974

Printed in Great Britain by Unwin Brothers Limited
The Gresham Press, Old Woking, Surrey
A member of the Staples Printing Group

ISBN 0 85047 204 0

8·75·6

Contents

How to Use Course Companions

Course Companions have been prepared by teachers who have detailed and first-hand experience of examinations and examining. They direct your attention to those areas of the subject which matter most, and which frequently cause confusion and difficulties to students.

Studying involves far more than merely sitting down and reading a book. Real study calls for much more active participation on the part of the student. Research shows that regular study and frequent revision are of far greater value in learning than last minute cramming. It is essential to work steadily throughout your course, making sure that you understand each topic before moving on to the next. To get the most out of these books you should first read that part which is relevant to the particular topic you are studying. Making your own notes and diagrams can be of considerable assistance in re-inforcing what you read—the mere fact of writing being a tremendous aid to memory. Working through the printed examples, stage by stage, and trying other examples, perhaps from past examination papers, will also be extremely useful.

Concentrated attention is physically very tiring, and can therefore only be maintained for short periods of time. By emphasising vital parts of the syllabus, **Course Companions** make study time more effective.

Introduction

This books is intended primarily for a student wishing to revise for an "O" level or C.S.E. examination. The subject matter has been cut to the essential minimum required. Throughout the text various phrases and definitions have been **printed in bold type** and a list of important formulae is on the final two pages. These are the points that need careful attention and in many cases should be **committed to memory**.

At the end of each group of topics appear a number of questions. These are of two types: Short Answer Questions test that the student has understood the preceding text and can apply the formula given there. The Short Answer type questions are becoming increasingly popular with examination boards. Then follow the more conventional Long Answer Questions. The answers to the questions follow each set; outlines of the calculations to the long answer questions have been given in most cases. **The student is urged to complete his answer to each question before consulting the suggested one.**

In an examination a student should **show all necessary working** clearly and on the same page as the solution, preferably on the right-hand side of the page. **Delay working out intermediate steps** and substituting for π until it is clearly necessary.

Set your work out **neatly** and **give concise reasons** for the steps involved. Do not forget to give your final answer in the required **units** and to the required **degree of accuracy**. Having obtained an answer to a problem it is always advisable to do **a rough check**, either mentally or written, to ensure that the answer is sensible. If you find a train travelling between two stations in 15 seconds or the cost of a house to be 50p then you should look through your working again.

In the main the accepted SI units and abbreviations have been used.

Numbers

An **integer** is a whole number. 3, 67 and 3549 are integers.

A **product** is the answer to a multiplication sum. Thus 56 is the product of 7 and 8.

A **quotient** is the answer to a division sum. Thus the quotient of 18/3 is 6.

A **digit** is one of the figures which make up the system. In the binary system the digits are 0 and 1. In the denary (decimal) system the digits are 0, 1, 2, 3, 4, 5, 6, 7, 8 and 9. If one number divides exactly into a second number, the first is a **factor** of the second, and the second is a **multiple** of the first. Thus 4 is a factor of 36 and 36 is a multiple of 4.

A **prime number** is one which has 1 and itself as its only factors. 5, 11, 13 and 23 are prime numbers.

The **prime factors** of a number are the prime numbers whose product gives the number. The prime factors of 36 are $2 \times 2 \times 3 \times 3$.

The largest number which divides exactly into two or more numbers is called their **highest common factor** (H.C.F.). The H.C.F. of 24 and 36 is 12.

The smallest number which is exactly divisible by two or more numbers is called their **lowest common multiple** (L.C.M.). The L.C.M. of 24 and 36 is 72.

H.C.F.'s and L.C.M.'s may best be found using prime factors.

Example: Find the H.C.F. and the L.C.M. of 56, 70 and 98.

$56 = 2 \times 2 \times 2 \times 7$
$70 = 2 \times 5 \times 7$
$98 = 2 \times 7 \times 7$

The H.C.F. contains only those factors common to all the numbers, i.e. $2 \times 7 = \textbf{14}$.

The L.C.M. must contain every factor of all the numbers, i.e. $2 \times 2 \times 2 \times 5 \times 7 \times 7 = \textbf{1960}$.

Example: Find the least number which, when divided by 9, 21 or 28 leaves a remainder 7 in each case.

$9 = 3 \times 3$ the L.C.M. $= 2 \times 2 \times 3 \times 3 \times 7$
$21 = 3 \times 7$ $= 252$
$28 = 2 \times 2 \times 7$ therefore the required number $= 252 + 7$
                                            $= \textbf{259}$

Fractions

Fractions are made up of two numbers, the **numerator** on the top and the **denominator** on the bottom.

A **proper fraction** is one in which the numerator is less than the denominator. $\frac{2}{3}$ is a proper fraction.

An **improper fraction** is one in which the numerator is greater than the denominator. $\frac{22}{7}$ is an improper fraction.

A **mixed number** consists of an integer and a fraction. $7\frac{1}{2}$ is a mixed number.

The value of a fraction is not changed by multiplying or dividing both the numerator and the denominator by the same number. Thus $\frac{4}{6} = \frac{2}{3}$ and $\frac{5}{15} = \frac{1}{3}$.

ADDITION AND SUBTRACTION

Fractions can be **added** or **subtracted** only if their denominators are the same. This is arranged by finding the L.C.M. of the denominators (**Least Common Denominator**) and multiplying numerator and denominator by the number required to make the denominator equal to the L.C.D. Thus to evaluate $\frac{2}{3} + \frac{3}{4} - \frac{1}{6}$ we note that the L.C.D. is 12 and the expression is therefore equal to

$$\frac{2 \times 4}{3 \times 4} + \frac{3 \times 3}{4 \times 3} - \frac{1 \times 2}{6 \times 2} = \frac{8 + 9 - 2}{12} = \frac{15}{12} = \frac{5}{4}$$

When mixed numbers occur in an expression to be evaluated deal with the whole numbers first and proceed as follows:

$$4\frac{1}{2} - 1\frac{3}{4} + \frac{1}{5} = 3 + \frac{10 - 15 + 4}{20} = 2\frac{19}{20}$$

MULTIPLICATION OF FRACTIONS

Two or more fractions are **multiplied** together by dividing the product of their numerators by the product of their denominators, after having first cancelled any common factors in numerators and denominators. Note that mixed numbers must first be converted to improper fractions.

For example:

$$\frac{3}{7} \times 5\frac{1}{4} \times \frac{2}{5} = \frac{3}{7} \times \frac{21}{4} \times \frac{2}{5} = \frac{9}{10}$$

DIVISION OF FRACTIONS

To **divide** by a fraction, **multiply** by the **inverse** of the divisor. For example:

$$\frac{5}{8} \div \frac{3}{4} = \frac{5}{8} \times \frac{4}{3} = \frac{5}{6}; \quad \text{and} \quad 2\frac{1}{7} \div 1\frac{2}{3} = \frac{15}{7} \times \frac{3}{5} = \frac{9}{7} = 1\frac{2}{7}$$

Operations

Numbers are combined by operations, such as adding, subtracting, multiplying, dividing, squaring and taking the square root. When more than one operation is involved, brackets are sometimes used to indicate the order in which the operations should be carried out. **The contents of the brackets are always evaluated first.** The word "of" is sometimes used to signify multiplication.

Example: $4 \times (5-3) = 4 \times 2 = 8$

Example: $4 \times (3+2) = 4 \times 5 = 20$

Example: $(5-3) \times (2+7) = 2 \times 9 = 18$

Example: $\frac{2}{3}$ of $15 = \frac{2}{3} \times 15 = 10$

Care must be taken when using brackets that they are correctly positioned:

$$\frac{(4+6)^2}{5} = \frac{10^2}{5} = \frac{100}{5} = 20$$

$$\left(\frac{4+6}{5}\right)^2 = \left(\frac{10}{5}\right)^2 = 2^2 = 4$$

$$\frac{(4)^2 + (6)^2}{5} = \frac{16+36}{5} = \frac{52}{5} = 10\frac{2}{5}$$

QUESTIONS

In questions 1–10, are the statements true or false?

1. A proper fraction is one in which the numerator is less than the denominator.

2. A mixed number is one containing different digits.

3. When finding a common denominator, the H.C.F. of the denominators is found first.

4. 25 is the product of 2 and 5.

5. 11 is a factor of 132.

6. 27 is a prime number.

7. The prime factors of 45 are $3 \times 3 \times 5 \times 2$.

8. The two fractions $\frac{8}{9}$ and $\frac{4}{5}$ are equal.

9. $\dfrac{(4-2)}{6} + \left(\dfrac{2+1}{3}\right)^2 = \frac{4}{3}$

10. $\frac{5}{11} \div \frac{10}{6} = \frac{1}{4}$

11. What is the smallest integer by which 84672 must be multiplied to make the product (a) a perfect square, (b) a perfect cube?
Write down the square root of (a) and the cube root of (b).

12. Find (a) the H.C.F., (b) the L.C.M. of the three numbers 36, 54 and 126.

13. When a certain number is divided by 17 the quotient is 67 and the remainder is 11. What is the number?

14. What is the least number which leaves a remainder 5 when divided by each of 6, 8 and 14? What is the next smallest number which leaves a remainder 5 when divided by 6, 8 and 14?

15. Express $\frac{7}{9} \times (2\frac{3}{7} - 1\frac{2}{3})$ as a single fraction in its lowest terms.

8

16. Simplify $\dfrac{2\frac{3}{7} - 1\frac{5}{8}}{\frac{3}{7} \times \frac{5}{8}}$

17. Simplify $\dfrac{(2\frac{1}{4})^2 - (1\frac{1}{5})^2}{1\frac{1}{20}}$

18. Simplify $\frac{7}{8}$ of $1\frac{2}{3} \div (1\frac{1}{6} - \frac{3}{4})$

19. Express $(1\frac{4}{5} - 1\frac{1}{3}) \div (2\frac{1}{3})^2$ as a single fraction in its lowest terms.

20. Simplify $\dfrac{\frac{2}{3} - \frac{3}{4}}{\frac{5}{7} - \frac{7}{10}}$

21. A sum of money was divided among 1 $\quad$ men; the first received $\frac{1}{2}$ the sum, the second $\frac{1}{8}$ and the rest each $\quad$ ceived £25. What was the sum?

22. One-third of a man's income was paid in taxes; of the remainder 7/10ths was spent on necessities, 1/8th on luxuries, 1/16th was given to deserving causes and the remainder, £150, was saved. What was his income?

ANSWERS

1t, 2f, 3f, 4f, 5t, 6f, 7f, 8f, 9t, 10f.

11. $2^3 = 8$
     $2^3 = 8$
     $3^2 = 9$
     3
     7^2

| 84672 |
| 10584 |
| 1323 |
| 147 |
| 49 |
| 1 |

(**Note.** First express 84672 in its prime factors. For a perfect square all indices must be even and for a perfect cube all indices must be multiples of 3.)

$\therefore 84\,672 = 2^6 \times 3^3 \times 7^2$

Ans. (a) 3, (b) 7; $2^3 \times 3^2 \times 7 = 504$; $2^2 \times 3 \times 7 = 84$

12. $36 = 2^2 \times 3^2$
    $54 = 2 \times 3^3$
    $126 = 2 \times 3^2 \times 7$
    (a) H.C.F. $= 2 \times 3^2 = 18$;
    (b) L.C.M. $= 2^2 \times 3^3 \times 7 = 756$

13. The number $= (17 \times 67) + 11 = 1150$

14. The L.C.M. is the least number which can be divided exactly by 6, 8 and 14. $\therefore$ We require the L.C.M. plus 5 and twice the L.C.M. $+ 5$. Answers. 173, 341.

Notes. 1. Be warned against false cancelling. Only common factors may be cancelled.

2. In division of fractions it is the fraction following the division sign which is inverted.

3. In addition and subtraction deal with any whole numbers first.

4. In multiplication and division mixed numbers are first changed into improper fractions.

5. When cancelling cross the cancelled figures carefully so that they can still be read. This will allow you to check your working and allow an examiner to see your method.

15. $\dfrac{7}{9} \times \left(2\dfrac{3}{7} - 1\dfrac{2}{5}\right) = \dfrac{7}{9}\left(1\dfrac{15-14}{35}\right) = \dfrac{7}{9}\left(1\dfrac{1}{35}\right)$

$= \dfrac{7}{9} \times \dfrac{36}{35} = \dfrac{4}{5}$

16. Exp^n. $= \dfrac{1 + \dfrac{24 - 35}{56}}{\dfrac{3}{7} \times \dfrac{5}{8}} = \dfrac{\dfrac{80 - 35}{56}}{\dfrac{3}{7} \times \dfrac{5}{8}} = \dfrac{45}{56} \times \dfrac{7}{3} \times \dfrac{8}{5} = 3$

(**Note.** When the difference of two squares occurs it is usually best to use the formula $a^2 - b^2 = (a - b)(a + b)$.)

17. **Method 1**

$$\text{Exp}^n. = \left(\frac{81}{16} - \frac{36}{25}\right) \div \frac{21}{20}$$

$$= \frac{2025 - 576}{16 \times 25} \div \frac{21}{20}$$

$$= \frac{1449}{16 \times 25} \times \frac{20}{21} = 3\tfrac{9}{20}$$

Method 2

$$\text{Exp}^n. = \frac{\left(2\tfrac{1}{4} - 1\tfrac{1}{5}\right)\left(2\tfrac{1}{4} + 1\tfrac{1}{5}\right)}{1\tfrac{1}{20}}$$

$$= \frac{1\tfrac{1}{20}\left(3\tfrac{9}{20}\right)}{1\tfrac{1}{20}}$$

$$= 3\tfrac{9}{20}$$

18. $\text{Exp}^n. = \dfrac{7}{8} \times \dfrac{5}{3} \div \dfrac{(14-9)}{12} = \dfrac{7}{\overset{}{\underset{2}{8}}} \times \dfrac{\overset{1}{5}}{\overset{}{\underset{1}{3}}} \times \dfrac{\overset{4}{12}}{\overset{}{\underset{1}{5}}} = 3\tfrac{1}{2}$

19. $\text{Exp}^n. = \dfrac{12-5}{15} \div \dfrac{49}{9} = \dfrac{\overset{1}{7}}{\underset{5}{15}} \times \dfrac{\overset{3}{9}}{\underset{7}{49}} = \dfrac{3}{35}$

20. $\text{Exp}^n. = \dfrac{\dfrac{8-9}{12}}{\dfrac{50-49}{70}} = -\dfrac{1}{12} \div \dfrac{1}{70} = -5\tfrac{5}{6}$

21. The first two men received $\frac{1}{2} + \frac{1}{8} = \frac{5}{8}$ of the sum.

∴ $\frac{3}{8}$ of the sum was the amount received by the 9 other men, i.e.
$9 \times £25 = £225$

∴ The sum was $£225 \times \dfrac{8}{3} = £600$

22. The fraction of his income paid away was

$$\frac{1}{3} + \left(\frac{7}{10} + \frac{1}{8} + \frac{1}{16}\right) \text{ of } \frac{2}{3} = \frac{1}{3} + \frac{7}{15} + \frac{1}{12} + \frac{1}{24}$$

$$= \frac{40 + 56 + 10 + 5}{120} = \frac{111}{120}$$

∴ $\dfrac{9}{120}$th of income $= £150$ ∴ Income $= \dfrac{£150 \times 120}{9} = £2,000$

Decimals

We express whole numbers by a system based on powers of 10. Thus 403 means 4 hundreds, no tens and 3 units, the position of each digit indicating its value. In the same way numbers less than unity are expressed as fractions with powers of 10 for their denominators, and known as **decimal fractions.** They are written following a decimal point, an extension of digital value by position. So 403·75 means 4 hundreds, no tens, 3 units, 7 tenths and 5 hundredths, or in figures,

$$403 \cdot 75 = 403 + \frac{7}{10} + \frac{5}{100} = 403 + \frac{75}{100} = 403\tfrac{3}{4}$$

Addition and subtraction

The addition and subtraction of decimals is merely an extension of the method used for integers (whole numbers). The decimal points should be written underneath one another.

Multiplication of decimals

Multiplication by 10 is simply a matter of moving the decimal point one place to the **right,** since hundredths become tenths, tenths become units and so on.

$$22 \cdot 34 \times 16 \cdot 4 = \frac{2234}{100} \times \frac{164}{10}$$
$$= \frac{2234 \times 164}{1000}$$
$$- 366 \cdot 376$$

$$
\begin{array}{r}
2234 \\
164 \\
\hline
2234 \\
13404 \\
8936 \\
\hline
366376 \\
\end{array}
$$

The above example illustrates the rule for multiplying two decimal numbers.

Rule 1. Multiply the numbers ignoring the decimal points.
     2. Add together the numbers of decimal places in the two numbers.
     3. Counting from the right, mark off this number of places in the product by inserting the decimal point.

Note. If the product ends with zeros, these must not be deleted until **after** the decimal point has been inserted.

Examples. $0 \cdot 2 \quad \times 0 \cdot 3 = 0 \cdot 06.$ $(2 \times 3 = 6)$
              $0 \cdot 065 \times 40 = 2 \cdot 6$ $(65 \times 40 = 2600)$

Short Methods of Multiplication

To multiply by:

5, move dec. pt. 1 place to the right and divide by 2;

25, move dec. pt. 2 places to the right and divide by 4;

125, move dec. pt. 3 places to the right and divide by 8.

DIVISION OF DECIMALS

Short division by an integer less than or equal to ($\leq$) 12 presents no difficulty:

$$71\cdot64 \div 4 = 17\cdot91 \qquad 4 \mid \underline{71\cdot64} \qquad\qquad 12 \mid \underline{263\cdot208}$$
$$263\cdot208 \div 12 = 21\cdot934 \qquad\qquad 17\cdot91 \qquad\qquad\qquad 21\cdot934$$

Division by 10 is simply a matter of moving the decimal point one place to the **left,** since units become tenths and so on.

Division by a **multiple of 10** should be done by factors, the division by the factor 10 being indicated by small marks to the left of the decimal point and the zero.

$$2\,0 \mid 18\,1\cdot62$$
$$9\cdot081$$

DIVISION BY A DECIMAL NUMBER

The method to be recommended is to make the **divisor a whole number.** Write the numbers as a fraction with the decimal points under one another. A line drawn after the last figure of the divisor will then indicate the positions of the points after the divisor has been made a whole number (i.e. by multiplying top and bottom by a power of 10). In the working write the first digit of the quotient over the last digit of the dividend used in step one of the division. The decimal point of the quotient will then come over that in the dividend.

$$21\cdot5688 \div 0\cdot264 = \frac{21\cdot568 \mid 8}{0\cdot264 \mid}$$

$$= \frac{21\,568\cdot8}{264}$$

$$= 81\cdot7$$

And, $21\cdot5688 \div 0\cdot00264 = \dfrac{21\cdot5688}{0\cdot00264}$

$$= \frac{2\,156\,880}{264}$$

$$= 8170$$

$$
\begin{array}{r}
81\cdot7 \\
264 \mid \overline{21568\cdot8} \\
2112 \\
\hline
448 \\
264 \\
\hline
1848 \\
1848 \\
\hline
\\
\cdots\cdots \\
\hline
\end{array}
$$

13

Also, $0.0215688 \div 26.4 = \dfrac{0.0 \mid 215688}{26.4 \mid} = \dfrac{0.215688}{264} = 0.000817$

Short Methods of Division
To divide by:
  5, move dec. pt. 1 place to the left and multiply by 2;
 25, move dec. pt. 2 places to the left and multiply by 4;
125, move dec. pt. 3 places to the left and multiply by 8.
To multiply a two-digit number by 11 write the sum of the digits between them (carrying 1 if necessary). For example:
$35 \times 11 = 385$; $56 \times 11 = 616$

DECIMAL PLACES AND SIGNIFICANT FIGURES

$22 \div 7 = 3.142857 \ldots$ etc. There are two ways of expressing this number more **approximately,** but nevertheless more usefully, since it would be extremely tedious if we had to multiply or divide by a number with several places of decimal.
The first way is to write the number **correct to so many decimal places.**
Correct to 1 decimal place it is 3.1
Correct to 2 decimal places it is 3.14
Correct to 3 decimal places it is 3.143
Note that in the last case the final figure has been "rounded up" to a 3, and not written as 2. This is done whenever the next figure is 5, 6, 7, 8 or 9.

The second method is to write the number so that it contains so many **significant figures.** Significant figures include **all** the figures in the number and not just those to the right of the decimal point.
Correct to 1 significant figure (1 s.f.) it is 3
Correct to 2 s.f.'s it is 3.1
Correct to 3 s.f.'s it is 3.14
Correct to 4 s.f.'s it is 3.143

The numbers 2704, 270.400 and 0.002704 all have **four** significant figures. The sequence of digits 2, 7, 0 and 4 is the same in each case. The other 0's alter the **place value** of the sequence and hence the magnitude of the number. When numbers are correct to a lower number of significant figures it is essential to retain the correct place value, e.g. 176250 correct to 2 s.f.'s is 180000, and to 1 s.f. it is 200000.

To obtain the answer to a problem correct to 3 s.f.'s it is important to retain exact values, or 4 figure accuracy if tables are used, throughout the work, and not to approximate until the final answer is reached.

When a division is carried out for an answer correct to say 5 s.f.'s the sixth figure should be calculated in order that the answer can be corrected as requested.

Indices

64 can be written as 8^2 where 8 is the **base** and 2 the **index.** Similarly $64 = 4^3$ or 2^6.

25 000 is the same as 2500×10 or 250×10^2 or 25×10^3.

Numbers in this form which have the **same base** may be combined together using the three basic rules of indices. These may be stated in general form:

$$a^m \times a^n = a^{m+n}$$
$$a^m \div a^n = a^{m-n}$$
$$(a^m)^n = a^{m \times n}$$

Examples:

$$8^3 \times 8^2 = 8^5$$
$$5^5 \div 5^3 = 5^2$$
$$(3^2)^4 = 3^8$$

The same rules apply to negative indices.

EXAMPLE

$$10^3 \div 10^5 = 10^{-2}$$

Note that

$$10^{-2} = \frac{1}{10^2} = \frac{1}{100} = 0 \cdot 01$$

and that

$$2 \cdot 5 \times 10^{-4} = 0 \cdot 000 25$$

15

Fractional indices are used to indicate the roots of numbers.

Thus $4^{\frac{1}{2}} = \sqrt{4} = 2$ and $27^{\frac{1}{3}}$ means the cube root of 27.

Fractional indices also obey the rules given above.

Examples:
$$(2^{\frac{1}{2}})^6 = 2^3$$
$$2^{\frac{1}{2}} \times 2^{\frac{1}{2}} \times 2^2 = 2^3$$

STANDARD FORM

It is convenient to write very large and very small numbers in standard form, i.e. as **A × 10^n** where **A** lies between 1 and 10, and **n** is an integer. For large numbers n is positive and for small numbers it is negative.

Example: $28\,300\,000 = 2{\cdot}83 \times 10\,000\,000 = 2{\cdot}83 \times 10^7$

Example: $0{\cdot}000\,043 = 4{\cdot}3 \div 100\,000 = 4{\cdot}3 \times 10^{-5}$

Problems are simplified when standard form is used throughout.

Example: $\dfrac{9 \times 10^8 \times 8 \times 10^{-4}}{4 \times 10^6} = 18 \times 10^{-2} = 1{\cdot}8 \times 10^{-1}$

Standard form has another advantage when using logarithms. When a number is written in standard form the index n is the same as the characteristic of the logarithm.

Examples: The logarithm of 2 is $0{\cdot}3010$

The logarithm of 2×10^6 is $6{\cdot}3010$

The logarithm of 2×10^{-9} is $\bar{9}{\cdot}3010$

QUESTIONS

In questions 1–4 are the statements true or false?

1. A decimal fraction is one whose denominator is a power of ten.

2. 29·067 written to three significant figures is 29·0.

3. 29·067 written to two decimal places is 29·06.

4. 0·0045 expressed in standard form is $4·5 \times 10^{-3}$

5. Select (a)–(e) to match, or have the same values as 1–5.

1. 3/5		(a)	0·64
2. 13/20		(b)	0·60
3. 16/25		(c)	0·63
4. 31/50		(d)	0·62
5. 63/100		(e)	0·65

6. Select (a)–(e) to match, or have the same values as 1–5

1. 0·05		(a)	5×10^{-2}
2. 1/2·5		(b)	$2·5 \times 10^{2}$
3. 500		(c)	5×10^{2}
4. 250		(d)	4×10^{-1}
5. 2·5		(e)	$2·5 \times 10^{0}$

7. Select (a)–(e) to match, or have the same values as 1–5

1. 1·2 × 1·4		(a)	$1·68 \times 10^{-1}$
2. 12 × 0·014		(b)	$1·4 \times 10^{1}$
3. 140 × 12		(c)	$1·68 \times 10^{0}$
4. $(1·2)^2$		(d)	$1·44 \times 10^{0}$
5. 1·96/0·14		(e)	$1·68 \times 10^{3}$

8. Simplify $\dfrac{3·6 \times 0·9}{0·75}$

9. Evaluate $\dfrac{18 \times (0·3)^2}{0·048}$

10. Find the exact value of $\dfrac{0.374 \times 2.63}{0.935}$

11. Evaluate $\dfrac{0.62 + 0.82}{0.3(7.62 - 4.42)}$ exactly.

12. Without the use of tables find the value of
$\dfrac{17.53 \times 0.106}{0.267}$, giving your answer correct to
(a) 3 decimal places, (b) 2 significant figures.

$\dfrac{3.142(5.4^2 - 4.6^2)}{10.64}$ correct to
(a) one place of decimals, (b) 3 significant figures.

14. Find without using tables, the value of
$\dfrac{(0.6507 \times 35.7) + (0.6507 \times 64.3)}{0.482(2.73 - 0.73)}$

15. Use the formula $s = \frac{1}{2}(u + v)t$ to find the value of s when $u = 4.63$, $v = 9.47$, $t = 2.4$.

16. Divide 1451·346 by 0·078, giving the answer (a) exactly, (b) correct to 4 significant figures.

17. Simplify $\dfrac{0.6}{2.5} + \dfrac{4.7}{1.25}$

18. Find the value of $\dfrac{3.44 + 2.09 - 1.26}{1.27 - 1.148}$

19. (i) Evaluate, without using tables, $\dfrac{36.7 \times 0.0784}{1.865}$, correct to 3 significant figures.
(ii) Use part (i) to write down the value of
(a) $\dfrac{3.67 \times 7.84}{186.5}$, (b) $\dfrac{0.367 \times 784}{0.1865}$

Notes
1. It is always advisable to do a rough check (R.C.), either mentally or written, to ensure that the decimal point is correctly placed.

2. If an **exact** answer is demanded, logarithms must not be used.

ANSWERS

1t, 2f, 3f and 4t.

Matching 5. 1(b), 2(e), 3(a), 4(d), 5(c).

6. 1(a), 2(d), 3(c), 4(b), 5(e).

7. 1(c), 2(a), 3(e), 4(d), 5(b).

8. R.C. $= \dfrac{4 \times 0 \cdot 9}{0 \cdot 8} = \dfrac{36}{8} \approx 4 \cdot 5$

$$\frac{3 \cdot 6 \times 0 \cdot 9}{0 \cdot 75} = \frac{36 \times \overset{3}{\cancel{9}}}{\underset{25}{\cancel{75}}} = \frac{108}{25} = 4 \cdot 32$$

9. R.C. $= \dfrac{18 \times 0 \cdot 09}{0 \cdot 05} = \dfrac{18 \times 9}{5} \approx 30$

$$\frac{18 \times (0 \cdot 3)^2}{0 \cdot 048} = \frac{18 \times 0 \cdot 09}{0 \cdot 048} = \frac{\overset{3}{\cancel{18}} \times \overset{45}{\cancel{90}}}{\underset{\underset{4}{\cancel{8}}}{\cancel{48}}} = \frac{135}{4} = 33 \cdot 75$$

10. R.C. $= \dfrac{0 \cdot 4 \times 3}{1} = 1 \cdot 2$

$$\frac{0 \cdot 374 \times 2 \cdot 63}{0 \cdot 935} = \frac{\overset{2}{\overset{\cancel{34}}{\cancel{374}}} \times 2 \cdot 63}{\underset{\underset{5}{\cancel{85}}}{\cancel{935}}} = \frac{5 \cdot 26}{5} = 1 \cdot 052$$

11. $\dfrac{0 \cdot 62 + 0 \cdot 82}{0 \cdot 3(7 \cdot 62 - 4 \cdot 42)} = \dfrac{1 \cdot 44}{0 \cdot 3 \times 3 \cdot 2} = 1 \cdot 5$

12. R.C. $\simeq \dfrac{20 \times 0\cdot 1}{0\cdot 3} = \dfrac{20}{3} \simeq 7$

$17\cdot 53 \times 0\cdot 106 = 1\cdot 85818$

$\dfrac{1\cdot 858 \mid 18}{0\cdot 267 \mid} = \dfrac{1858\cdot 18}{267}$

$$
\begin{array}{r}
1753 \\
106 \\
\hline
1753 \\
10518 \\
\hline
185818
\end{array}
$$

$$
\begin{array}{r}
6\cdot 9594 \\
267)\overline{1858\cdot 18} \\
1602 \\
\hline
2561 \\
2403 \\
\hline
1588 \\
1335 \\
\hline
2530 \\
2403 \\
\hline
1270 \\
1068 \\
\hline
202 \\
\end{array}
$$

Answer (a) 6·959
       (b) 7·0

Note. It is essential to put in the zero for (b).

13. $\dfrac{3\cdot 142(5\cdot 4^2 - 4\cdot 6^2)}{10\cdot 64} = \dfrac{3\cdot 142 \times 0\cdot 8 \times 10}{10\cdot 64}$ (using the difference of two squares)

$= \dfrac{25\cdot 136}{10\cdot 64}$

$= 2\cdot 362$ (by long division)

Answer (a) 2·4   (b) 2·36

14. Expression $= \dfrac{0\cdot 6507(35\cdot 7 + 64\cdot 3)}{0\cdot 482 \times 2}$ (common factor 0·6507)

$= \dfrac{0\cdot 6507 \times 100}{0\cdot 964}$

$= \dfrac{65\cdot 07}{0\cdot 964} = 67\cdot 5$ exactly (by long division)

Note. If no degree of accuracy is asked for in the question, it is reasonable to assume that the answer will be exact.

15. $s = \dfrac{4\cdot63 + 9\cdot47}{2} \times 2\cdot4 = 14\cdot10 \times 1\cdot2 = 16\cdot92$

16. R.C. $\simeq \dfrac{1500}{0\cdot08} = \dfrac{150\,000}{8} \simeq 20\,000$

$\dfrac{1451\cdot346}{0\cdot078} = \dfrac{1\,451\,346}{78}$

$$
\begin{array}{r}
18607 \\
78)\overline{1451346} \\
\underline{78} \\
671 \\
\underline{624} \\
473 \\
\underline{468} \\
546 \\
\underline{546} \\
\cdots
\end{array}
$$

Answer (a) 18607
        (b) 18610

17. $\dfrac{0\cdot6}{2\cdot5} + \dfrac{4\cdot7}{1\cdot25} = \dfrac{6}{25} + \dfrac{470}{125} = \dfrac{6}{25} + \dfrac{94}{25} = \dfrac{100}{25} = 4$

18. $\dfrac{3\cdot44 + 2\cdot09 - 1\cdot26}{1\cdot27 - 1\cdot148} = \dfrac{4\cdot27}{0\cdot122} = 35$

19. R.C. $= \dfrac{40 \times 0\cdot08}{2} = 1\cdot6$

$$
\begin{array}{r}
367 \\
784 \\
2569 \\
2936 \\
1468 \\
\overline{287728}
\end{array}
\qquad
\begin{array}{r}
1\cdot542 \\
1865)\overline{2877\cdot28} \\
\underline{1865} \\
10122 \\
\underline{9325} \\
7978 \\
\underline{7460} \\
5180 \\
\underline{3730} \\
1450
\end{array}
$$

$\dfrac{2\cdot877\ |\ 28}{1\cdot865\ |} = \dfrac{2877\cdot28}{1865}$

(i) Answer 1·54 (to 3 s.f.)

(ii) (a) R.C. $= \dfrac{4 \times 8}{200} = 0\cdot16$

      Answer 0·154 (to 3 s.f.)

   (b) R.C. $= \dfrac{0\cdot4 \times 800}{0\cdot2} = 1600$

      Answer 1540 (to 3 s.f.)

Ratios

A gardening book gives the basic recipe for mixing concrete for paths as 4 parts shingle, 2 parts sand and 1 part cement. The ingredients should therefore be mixed in the ratio **4:2:1**. The actual amounts of material used will depend on the size of the path to be constructed. For a small job it may be sufficient to use 4 bucketfuls of shingle, 2 of sand and 1 of cement. For a larger path 12 barrowloads of shingle, 6 of sand and 3 of cement may be required. Both mixtures would, however, contain the materials in the same ratio.

For every 7 parts of concrete (i.e. $4+2+1$) four parts are shingle. 1 m³ of concrete contains 4/7 m³ of shingle, 2/7 m³ of sand and 1/7 m³ of cement.

Ratios should be expressed as simply as possible. 12:4, for example, is more simply expressed as 3:1, and $2\frac{1}{2}:1\frac{1}{4}$ is better given as 2:1. Ratios may be given in fractional form. For example a ratio of 4:5 means that the first part is 4/5 of the second, and the second is 5/4 of the first.

Example: Two men's wages are in the ratio 4:5. If the first earns £20 find the wages of the second.

The second man's wages are 5/4ths of the first, i.e. £20 × 5/4.

To increase 120 in the ratio 3:2 means to multiply 120 by 3/2.
To decrease 120 in the ratio 2:3 means to multiply 120 by 2/3.

Example: A, B and C hold 48, 120 and 72 Premium Bonds respectively. They agree to pool their bonds, so that any prize money is distributed in the ratio of their holdings. If they win £100 how much does each receive?

Ratio of holdings = 48:120:72 = 2:5:3
The sum of these three numbers is 10, so:
A receives 2/10ths of £100 = £20
B receives 5/10ths of £100 = £50
C receives 3/10ths of £100 = £30

THE SCALE OF MAPS

The scale of maps may be expressed as a ratio or as a **Representative Fraction** (R.F.), for example as $1:25000$ or $1/25000$.

It might also be given as a direct comparison of distances, for example as 1 cm representing 1 km. This may be converted to an R.F., but care must be taken with units.

$1 \text{ km} = 1000 \text{ m} = 100000 \text{ cm}$

$\therefore$ 1 cm to the kilometre $= 1:100000$ or $1/100000$

The scale is necessary to convert actual distances to map measurements and vice versa.

Example: The scale of a map is $1:25000$.
 (i) Find the actual length of a path which, on the map, is 1·2 cm long.
(ii) Find the length on the map of a road which is 2 km long.

Length of path $= 1·2 \times 25000$ **cm** $= 300$ **m**

Length of road $= 2 \times 1/25000$ **km** $= 8$ **cm**

When areas are involved the R.F. must be squared.

Example: If a field on the above map occupies 4 cm^2 find its true area.

Area $= 4 \times 25000 \times 25000$ **cm^2**

$= 4 \times \frac{1}{4} \times \frac{1}{4}$ km$^2 = \frac{1}{4}$ **km^2**

MODELS

A scale model has all its dimensions reduced from the original in the same ratio.

Example: A model barn, made to a scale of $1:40$, has dimensions 50 cm by 30 cm by 20 cm, and is in the shape of a rectangle. What is the volume of the real barn?

Volume $= 50 \times 40 \times 30 \times 40 \times 20 \times 40$ **cm^3**

$= 1920$ **m^3**

When dealing with maps and models great care must be taken with units.

PROPORTION

For a particular substance the ratio **mass:volume** is constant and is known as the **density** of the substance.

$$\text{Density} = \frac{\text{Mass}}{\text{Volume}}$$

Another way of expressing this relationship is to say that **mass is directly proportional to volume.** A graph of mass against volume would give **a straight line through the origin.**

If 4 oranges cost 10 p then the cost of 7 oranges is equal to the cost of 4 oranges **increased in the ratio 7:4,** i.e. 10 p × 7/4 = $17\frac{1}{2}$ p. This is another case of direct proportion—the **more** oranges bought the **more** the cost; the **fewer** oranges bought the **less** the cost would be.

Next consider the time taken by different numbers of men to complete a certain job, assuming that the men all work at the same rate. Clearly the **more** men at work the **shorter** the time taken, and the **fewer** the men at work the **longer** the time. This is a case of **inverse proportion,** the time being **inversely proportional** to the number of men.

Example: If 15 men do a job in 8 days how long will 12 men take to do the same job?
The number of men is decreased in the ratio 15:12 or 5:4. Since this is a case of inverse proportion, the time is increased in the ratio 5:4.
Time = 8 × 5/4 = **10 days**

The above examples illustrate simple proportion. The method can be extended to **compound proportion.**

Example: If 15 men earn £1 080 in 3 weeks how long will it take 8 men to earn £864?
Here the times are directly proportional to the amounts earned and inversely proportional to the number of men. The time is increased because there are fewer men and decreased because the total money earned is less.

8 men earn £864 in $3 \times \frac{15}{8} \times \frac{864}{1080}$ = **$4\frac{1}{2}$ weeks**

PERCENTAGES

A percentage is a fraction with denominator 100.
For example

$$\frac{1}{2} = \frac{50}{100} = 50\%$$

To express any fraction or decimal as a percentage multiply by 100
and write % after it.

For example

$$\frac{3}{16} = \frac{3}{16} \times 100\% = 18\frac{3}{4}\% \text{ or } 18 \cdot 75\%$$

$$0 \cdot 625 = 0 \cdot 625 \times 100\% = 62 \cdot 5\%$$

$$4\% \text{ of } £8 \text{ means } \frac{4}{100} \times £8 = £\frac{32}{100} = £0 \cdot 32$$

To express one quantity as a percentage of another first express
the one as a fraction of the other and then convert to a percentage.

Example: Express £3·50 as a percentage of £40.

Answer is $\frac{3 \cdot 50}{40} \times 100\% = 8\frac{3}{4}\%$

PERCENTAGE CHANGES

An increase of $r\%$ means that 100 becomes $100 + r$. Therefore a
number N increases in the same ratio, i.e. $(100 + r) : 100$

$$N \text{ increases to } N \times \frac{100 + r}{100}$$

Similarly a decrease of $r\%$ means that 100 becomes $100 - r$ and the
change is in the ratio $(100 - r) : 100$

$$N \text{ decreases to } N \times \frac{100 - r}{100}$$

QUESTIONS

Mark questions 1–4 true or false.

1. The ratio 21:7 is the same as 3:1.

2. 250 increased in the ratio 6:5 becomes 300.

3. A map scale where 1 cm represents 5 km has an R.F. of 1/100 000.

4. A model has dimensions 1/20 of the object.
   A volume of 4 cm³ on the model represents 1600 cm³.

5. Select (a)–(e) to match, or have the same value as 1–5.
   1. Increase £1.20 in the ratio 3:2   (a) £1.20
   2. Decrease £1.20 in the ratio 2:3   (b) £0·72
   3. Increase 60 p by 20%   (c) £0·765
   4. Decrease 90 p by 15%   (d) £0·80
   5. Increase £0·90 by 1/3   (e) £1·80

In questions 6, 7 and 8 place a tick against the only correct answer.

6. The ratio 35 cm per second:500 metres per minute, expressed in its simplest form is:
   (a) 105:500   (b) 21:100   (c) 35:25
   (d) 21:500   (e) 7:5

7. A faulty milometer records 36 miles for a 42 mile journey. It therefore records on a 63 mile journey:
   (a) 45 miles   (b) 54 miles   (c) 73·5 miles
   (d) 57 miles   (e) 48 miles

8. Twelve men can do a piece of work in 15 days. It will therefore take 20 men:
   (a) 25 days   (b) 32 days   (c) 9 days
   (d) 16 days   (e) 4·5 days

9. Calculate the area, in square metres, on the ground, represented by an area of 72 cm² on a plan drawn to a scale of 1:1000 (i.e. R.F. = 1/1000).

10. A fertiliser for a lawn was prepared by mixing three constituents A, B and C in the ratio $5:3:2$ by weight. 10 kg of A costs 60 p, 10 kg of B costs 40 p, and 10 kg of C costs 35 p.
What was the cost of each constituent in 1000 kg of the fertiliser?
If the fertiliser was spread at the rate of 0·05 kg per m^2, how many kg of B are used on a lawn measuring 20 m by 12 m?

11. The sides of a triangle are in the ratio $5:7:8$. If the longest side is 6·4 cm, what is the length of the perimeter?
The perimeter of a similar triangle is 28 cm. What is the length of its smallest side?
What is the ratio of the area of the first triangle to that of the second?

12. If 6 dogs can be kept for 10 days at a cost of £7.50, how many dogs can be kept at the same rate for 12 days at a cost of £18?

13. Three men, A, B and C invested £8 500, £6 250 and £1 750 respectively in forming a business. It was agreed that C should run the business and receive a salary of £1 500 p.a., and that the remainder of the net profit should be divided in the ratio of capital invested. What did each receive in a year when the net profit was £3 810?

14. A and B form a partnership in which A invested £5,600 and B invested £7 200. 12% of the total annual profit was used to expand the business, and A received an annual salary of £850. The remainder of the profit was shared between them in proportion to their invested capital. Find the total profit in a year when A received a total of £1 200.

ANSWERS

1t, 2t, 3f and 4f.

Matching (question 5): 1(e), 2(d), 3(b), 4(c), 5(a).

Correct choice: 6(d), 7(b), 8(c).

9. Linear Scale $1/10^3$, Area Scale $1/10^6$. 72 cm^2 represents $72 \times 10^6 \text{ cm}^2$

or $\dfrac{72 \times 10^6}{100 \times 100} \text{ m}^2 = 7200 \text{ m}^2$

10. 500 kg of A costs £30, 300 kg of B costs £12 and 200 kg of C costs £7. Area 240 m^2 requires 12 kg, 3/10ths of this is B, i.e. 3·6 kg.

11. $\dfrac{8}{20}$ of perimeter $= 6\cdot4$ ∴ perimeter $= \dfrac{6\cdot4 \times 20}{8} = 16$ cm

Smallest side $= \dfrac{5}{20} \times 28 = 7$ cm

Linear ratio $= 16:28 = 4:7$ ∴ Ratio of areas $= 4^2:7^2 = 16:49$

12. (Note. The number of dogs is directly proportional to the money available and inversely proportional to the times.)

£18 will last 12 days for $6 \times \dfrac{18}{7\frac{1}{2}} \times \dfrac{10}{12}$ dogs $= 12$ dogs

13. Total sum invested $=$ £16 500

∴ C receives £1 500 $+ \dfrac{1\,750}{16\,500} \times ($£3 810 $-$ £1 500$)$

$=$ £1 500 $+ 17\cdot5 \times \dfrac{\text{£2 310}}{165} = $ £1 500 $+$ £17·5 $\times 14 = $ £1 745

B receives £62·5 $\times 14 = $ £875,
and A receives £85 $\times 14 + $ £1 190

14. Total invested capital $=$ £12 800
In the "share out" A received £$(1\,200 - 850) = $ £350

∴ the shared profit $=$ £350 $\times \dfrac{12\,800}{5\,600} = $ £800

∴ 88% of the total profit $=$ £1 650

∴ Total profit $= £\dfrac{1\,650 \times 100}{88} = $ £1 875

Profit and Loss

Unless otherwise stated percentage profit and loss are considered **with respect to the cost price** (C.P.). Therefore when the gain is $r\%$ the profit is $r/100 \times$ C.P. and the selling price (S.P.) is the C.P. increased by $r\%$.

Thus S.P. $= \dfrac{100+r}{100} \times$ C.P.

Inversely, the C.P. $= \dfrac{100}{100+r} \times$ S.P.

Examples:

1. C.P. is £1·50, percentage profit is 20%.
   Calculate profit and S.P.

   Profit $= £1·50 \times \dfrac{20}{100} = 30$ p

   S.P. $=$ C.P. $+ 30$ p $= £1.80$

2. C.P. is £1·50. Percentage loss is 10%.
   Calculate loss and S.P.

   Loss $= £1·50 \times \dfrac{10}{100} = 15$ p

   S.P. $=$ C.P. $- 15$ p $= £1·35$

3. S.P. is £2·40, percentage profit is 25%.
   Calculate C.P. and profit.

   C.P. $= \dfrac{100}{100+25} \times £2·40 = £1·92$

   Profit $= £2·40 - £1·92 = 48$ p

4. S.P. is £6·50, C.P. is £5·00, calculate percentage profit.

   Profit $= £1·50$, percentage profit $= \dfrac{£1·50}{£5·00} \times 100\% = 30\%$

TABULATED FORM OF SOLUTION

In questions dealing with successive percentage changes the adoption of a tabular form for setting out the data and the use of **representative numbers** often simplifies the reasoning.

Example: If an article is sold for £40 the profit is 12%. For what must it be sold to make a profit of 26%?

	C.P.	1st S.P.	2nd S.P.
Representative numbers	100	112	126
Actual values		£40	£x

The **actual values are proportional to the representative numbers** and there is no need to calculate the cost price.

$$\frac{x}{40} = \frac{126}{112} \quad \therefore x = 40 \times \frac{126}{112} = 45$$

∴ the second selling price = £45.

Example: A manufacturer sold a car to a retailer for £640 at a profit of 24%; the retailer sold it to the first owner at a profit of 20%; later the first owner sold it to the second owner at a loss of 25%. What did the second owner pay?

	M's C.P.	R's C.P.	1st owner's C.P.	2nd owner's C.P.
Rep. numbers	100	124	$124 \times \frac{120}{100}$	$124 \times \frac{120}{100} \times \frac{75}{100}$
Actual values		£640		£x

Fill in the table as above. Always denote the **original value** (in this example the manufacturer's C.P.) by the **representative number 100.** There is no need to fill in the other two gaps. £x is £640 changed in the ratio of the two corresponding representative numbers.

$$\therefore \pounds x = \pounds 640 \times \frac{124 \times \dfrac{120}{100} \times \dfrac{75}{100}}{124}$$

$$= \pounds 576$$

Averages

In an examination there are 20 candidates. Their total marks added together are 840, thus the average mark is $840 \div 20 = 42$.
If there were another candidate who obtained 63 marks this would change the average thus:

$$\frac{840 + 63}{20 + 1} = \frac{903}{21} = 43$$

When averages are changed in this way it is important to work in terms of the totals, and not to attempt short cut methods.

Example: A cricketer has an average of 25 runs for 10 innings. In his next match he scores 36, what does his average become?

$$\text{Average} = \frac{\text{Total runs}}{\text{Total innings}} = \frac{(25 \times 10) + 36}{10 + 1} = \frac{286}{11} = 26$$

You must not take the average of his previous average and the next score. The average of 25 and 36 is 30·5 which is clearly wrong because it gives too much importance to the final score.

Example: The average height of a group of 5 boys is 130 cm. One more boy joins the group and the average becomes 127 cm. What is the height of this boy?

Total of height of first five boys $= 5 \times 130$ cm $= 650$ cm
Total of height of the six boys $\;= 6 \times 127$ cm $= 762$ cm

By subtraction, the height of the sixth boy is 112 cm.
Again note that the totals have been used to work out the answer.

AVERAGE SPEEDS

During a journey the speed of a car will change considerably from a maximum of possibly 100 km/hour to zero when the car is stuck in a traffic jam.

$$\text{Average speed} = \frac{\text{Total distance}}{\text{Total time}}$$

Graphs

Graphs are used to illustrate visually the relationship between two quantities.

General rules for plotting graphs are as follows:

1. Draw two perpendicular lines as axes.

2. On each axis **state the quantity** which is to be measured along it, and give it **units**. Thus an axis may be labelled "Time in seconds" or "Speed in km/hour".

3. If you are not told to use a particular scale choose one so that the graph almost **fills the page.** But at the same time choose a **convenient scale,** such as 1 large square for 5 units. Never choose a scale like 1 square for 3 units, this would make it difficult to plot points accurately.

4. Having chosen the scales for both axes, graduate each scale clearly.

5. Plot each point with a pencil "dot" and a small circle round it, or mark it with a small cross.

6. Always use a hard, sharp pencil both for plotting points and drawing in the line. Avoid drawing lines which waver about or become "fuzzy".

7. Show any necessary calculations and a table of values on a separate sheet of paper, unless you are given the table in the question.

8. If there are two or more graphs on the same axes then each line should be clearly labelled.

STRAIGHT LINE GRAPHS

If two quantities are **directly proportional** to each other their graph is a **straight line through the origin.** The extension of a spring, for example, varies directly as the load applied to it, so a graph of load against extension would be a straight line through the origin. (No load gives no extension.)

Strictly speaking, it is necessary to plot only two points in order to draw a straight line graph, and if one of these is the origin the task is very simple. But it is always advisable to plot one other point as a check.

CONVERSION GRAPHS

A straight line graph is often used as a means of converting units. The example below shows the conversion of £ sterling to U.S. dollars. We are told that £1 = £2·54 dollars, and of course we know that the graph will pass through the origin as £0 = 0 dollars.
As a check we can make use of the fact that the line passes through the point (£0·50, 1·27 dollars).

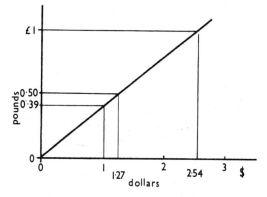

From the graph it is possible to convert quickly from £'s to dollars and vice versa. Obtaining information about points between the plotted ones is called **interpolation.**

33

To find how much English money is equivalent to 1 dollar, for example, we draw a vertical line from the 1 dollar mark on the horizontal axis until it cuts the line. A horizontal line is then drawn to cut the £ axis and the sterling value read off. In this case it is £0·39 to the nearest penny.

TRAVEL GRAPHS

If a man walks, or a vehicle is driven at a steady (constant) speed, then equal distances are travelled in equal intervals of time and a graph of distance against time is a straight line.

Speed, which is distance ÷ time, can be calculated from a distance–time graph by finding the **slope of the graph.** This is illustrated in the following example, which shows the journey of a coach tour.

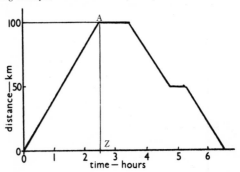

The horizontal portions of the graph indicate that the coach has stopped during these periods. The general shape of the graph shows that the coach travelled 100 km on the outward journey and then returned to its starting point.

The speed of the coach during the first part of the journey, represented on the graph from O to A, is given by the slope of OA, that is:

$$\text{Speed} = \frac{\text{Distance } AZ}{\text{Time } OZ} = \frac{100 \text{ km}}{2 \cdot 5 \text{ h}} = 40 \text{ km/h}$$

TRAVEL GRAPHS FOR TWO VEHICLES

If two vehicles travel between two points, A and B in opposite directions, then their travel graphs, drawn on the same axes, may be used to find where and when they meet. This is illustrated in the following example:

A and B are two towns 200 km apart. A passenger train leaves A at 9.00, travels 80 km in the first hour, stops for half and hour, and then completes its journey by 12.00.
A freight train leaves B at 10.00 and travels at a steady speed of 50 km/h towards A. Find when and where the trains pass.
To draw the graph for the freight train we must first find out what time it arrives at A.

$$\text{Time taken} = \frac{\text{distance}}{\text{speed}} = \frac{200 \text{ km}}{50 \text{ km/h}} = 4 \text{ h}$$

Thus the freight train arrives at 14.00.

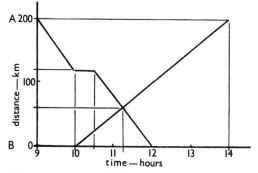

The trains pass at 11.15, 60 km from A

35

QUESTIONS

In Nos. 1–4 tick the only correct answer.

1. A car was bought by a garage for £240 and eventually sold at a loss of 5%. The selling price was:
    (a) £252   (b) £230   (c) £128
    (d) £216   (e) None of these

2. Mr Steel weighs 75 kg, Mrs Steel weighs 60 kg and Mark, their only son, weighs 40 kg. If the average weight of the family is 51 kg, find the weight of their only daughter. Is it:
    (a) 25 kg   (b) 31 kg   (c) 62 kg
    (d) 29 kg   (e) None of these?

3. A car started a journey of 480 km at 16.50 on 1st Sept. and completed it at 00 20 on 2nd Sept. The average speed of the car was:
    (a) 64 km/h   (b) 48 km/h   (c) 32 km/h
    (d) 24 km/h   (e) 30 km/h

4. A man saved 25 p by paying cash for an article marked at £5. His discount was:
    (a) 10%   (b) 50%   (c) 1%
    (d) 5 p in the £   (e) 10 p in the £

5. A house agent's fee for selling a house is 5% of the first £500 of the cost of the house and 2·5% of the remainder. What is the cost of the house if the agent's fee is £210?

6. An article was listed by the manufacturer at 20% above the cost of production. A discount of 4% off the list price was given to a dealer who re-sold the article for £18, thereby making a profit of 25% on his outlay. What did it cost the manufacturer to produce the article?

7. Two powdered substances, A and B, are mixed so that the mixture contains 44% by weight of A. What weight of A must be added to 75 kg of the mixture to make a new mixture containing 52% of A?

36

8. At an auction sale the average price of **6** articles was **75** p and that of **4** of them was **60** p. Calculate the prices of the other two if they differed by 12 p.

9. Tea costing **60** p per kg is blended with tea costing **70** p per kg. The blend is sold for **80** p per kg and this produces a profit of **20%**. Calculate the cost price of the mixture and the ratio in which the two teas are blended.

10. A boat travels a certain distance upstream at a speed of **7·5** knots and returns at **11·25** knots. Find its average speed for the double journey.

11. A cyclist left a town A at **9** a.m. to cycle to a town B, **15** miles away. Half an hour later a man set out to walk at a steady pace from B towards A. The cyclist travelled at a steady 8 mile/h until 10 o'clock, when he stopped for $\frac{1}{4}$ h and then proceeded at a reduced steady speed to arrive at B at **20** min past **11**. If the cyclist and pedestrian passed one another when they were **11** miles from A, find from the graph the time of meeting and the steady speed at which the pedestrian was walking and the reduced speed of the cyclist.

ANSWERS

1(e), 2(d), 3(a), 4(d).

5. Total fee = £210 and fee for 1st £500 = £25

∴ Fee for remainder of cost = £185

∴ Remainder of cost = £185 × $\dfrac{100}{2\frac{1}{2}}$ = £7 400

∴ Total cost = £7 900

6.

Cost of Production	List Price	Cost to Dealer	Selling Price
100	120	$120 \times \dfrac{96}{100}$	$120 \times \dfrac{96}{100} \times \dfrac{125}{100}$
£x			£18

$$\frac{x}{100} = \frac{18}{120 \times \dfrac{96}{100} \times \dfrac{125}{100}}$$

$$x = \frac{18 \times 100 \times 100 \times 100}{120 \times 96 \times 125} = 12\cdot5$$

∴ Cost of production is £12.50

Note. Neither the List Price nor the Cost to Dealer has to be calculated.

7. The first mixture contains $\dfrac{44}{100} \times 75$ kg of A.

If x kg are added, the second mixture contains

$\left(\dfrac{44}{100} \times 75 + x\right)$ kg of A and this is $\dfrac{52}{100}(75 + x)$ kg

∴ $\dfrac{44}{100} \times 75 + x = \dfrac{52}{100}(75 + x)$

∴ $825 + 25x = 975 + 13x$

∴ $\qquad 12x = 150$

∴ $\qquad x = 12\tfrac{1}{2}$

∴ $12\tfrac{1}{2}$ kg of A must be added

8. Total cost is $6 \times 75\,\text{p} = £4\cdot50$
Cost of four is $4 \times 60\,\text{p} = £2\cdot40$
Therefore cost of the other two = £2·10
One cost £1·05 + 6p and the other £1·05 − 6p, i.e. £1·11 and 99p.

9. Cost of mixture $= 80\,\mathrm{p} \times \dfrac{100}{120} = 66\tfrac{2}{3}\,\mathrm{p}$ per kg

	1st tea	2nd tea	mixture
cost per kg	60 p	70 p	$66\tfrac{2}{3}$ p
gain/loss per kg	$+6\tfrac{2}{3}$ p	$-3\tfrac{1}{3}$ p	—

These losses and gains must balance; this will occur if they are mixed in the ratio $1:2$.

10. Let the distance upstream be x nautical miles.
    Then the total distance $= 2x$ nautical miles,

and the time taken $= \dfrac{x}{7\frac{1}{2}} + \dfrac{x}{11\frac{1}{4}} = \dfrac{18\frac{3}{4}x}{7\frac{1}{2} \times 11\frac{1}{4}}$ h

∴ The average speed $= 2x \div \dfrac{75 \times 2 \times 4x}{4 \times 15 \times 45}$ knots $= 9$ knots

11. First draw the cyclist's graph OPQ and then join Q to R where R is 11.20 and 15 miles from A. A horizontal line 11 miles from A determines M, the point and time of meeting. SM is the travel graph for the pedestrian and its gradient gives his speed, and the gradient of QR gives the reduced speed of the cyclist, 6·5 mile/h.

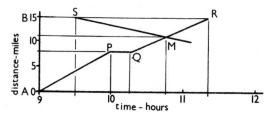

Hire Purchase

When buying an expensive item it may be more convenient to use hire purchase rather than making a single cash payment. A **deposit** is normally required. Often the law requires that this deposit be at least a certain percentage of the cash price. The actual percentage varies from item to item and from time to time as the government of the day wishes to encourage or discourage spending. The remainder of the money is paid by **instalments,** usually on a weekly or monthly basis. A purchaser using this system always pays more than the cash price for the article, since he is effectively borrowing money for the period of the transaction.

Questions set on hire purchase often ask you to find:
1. the extra amount paid as a result of the hire purchase,
2. the percentage of the cash price which this extra amount represents.

Example: A man buys a car on hire purchase, the cash price being £800. He pays a deposit of 20% and 24 monthly payments of £32·50.

He pays 20% of £800 deposit, that is $\frac{20}{100} \times £800 = £160$

plus the monthly instalments of $24 \times £32·50 = £780$

Total amount paid under the hire purchase $= £160 + £780 = £940$

Therefore the extra paid $= £940 - £800 = $ **£140**

As a percentage of the cash price this is $\frac{140}{800} \times 100\% = $ **$17\frac{1}{2}\%$**

If, for some reason, the instalments are not paid the shop or firm with whom the purchaser made the agreement can insist on the return of the goods.

A mortgage is a type of hire purchase used when buying property, such as a house. In this case the repayments may be spread over 25 or 30 years.

Gas and Electricity Charges

There are various ways by which Gas and Electricity Boards charge for their services. Typically, an electricity bill is made up of **a fixed charge** plus a **charge per unit** (kilowatt-hour) for electricity consumed during the period of the account. The bill usually shows the meter reading at the beginning and end of the period.

Example: During the period ended 21st March 1972 an electricity meter reading increased from 21 088 units to 22 508 units. Electricity is charged at 0·8p per unit plus a fixed charge of £1·72. Find the total bill for the period.

Units consumed $= 22\,508 - 21\,088 = 1420$
Cost at 0·8p per unit $= 1420 \times 0\cdot8\,\text{p} = £11\cdot36$
Total cost $= £11\cdot36 + £1\cdot72 = $ **£13·08**

To encourage people to use electricity during off-peak periods, usually during the night, it is available at a cheaper rate.

Gas boards encourage people to use more gas by having a **two-part tariff.** After using a fixed amount of as at a high cost the rest is paid at a lower rate.

Example: A householder may elect to be charged for gas in either of two ways:
1. First 4000 ft³ at 8·32p per 100 ft³
   Next 100 000 ft³ at 5·40p per 100 ft³
2. A fixed charge of £2·20 per quarter, and all gas at 4·82p per 100 ft³
If he uses 60 000 ft³ during a quarter year, which method is cheaper, and by how much?
Method 1: 4 000 ft³ at 8·32p per 100 ft³
$\quad\quad\quad = 40 \times 8\cdot32 = 332\cdot8\,\text{p} = £3\cdot328$
$\quad\quad\quad$ plus 56 000 ft³ at 5·40p per 100 ft³
$\quad\quad\quad = 560 \times 5\cdot40 = 3024\,\text{p} = £30\cdot24$
$\quad\quad\quad$ Total $= £3\cdot328 + £30\cdot24 = $ **£33·57** to nearest p.
Method 2: 60 000 ft³ at 4·82p per 100 ft³
$\quad\quad\quad = 600 \times 4\cdot82 = 2\,892\,\text{p} = £28\cdot92$
$\quad\quad\quad$ Total $= £2\cdot20 + £28\cdot92 = $ **£31·12**
The second method is cheaper by £2·45

Rates

Local Authorities, such as County Councils and Rural District Councils, raise money by levying rates. The money is used for a variety of purposes, for example:
County Council: education, fire service, police, highways.
Rural District Council: housing, sewerage, parks and pleasure grounds, refuse collection.

Every property, such as a house or a shop within the district is given a **rateable value.** This value depends on the type and size of the property and is expressed in £. The rateable value of a house might be £120 and that of a small factory £560.
The **general rate** is the amount that each property owner has to pay, every year, for each £ of rateable value. The rate is usually reduced for house owners. The general rate might be 70p and this may be reduced to 55p for a dwelling house.

Thus the householder pays $£120 \times \dfrac{55}{100} = £66$

and the factory owner $£560 \times \dfrac{70}{100} = £392$

The general rate might be divided up as follows:

Rural District purposes	14·7p
County purposes	50·3p
Main Drainage Authority	5·0p
	70·0p

The estimated production of a penny rate is used by the Council to calculate what rate to charge. If the rateable value of the council district is £564 000 then a penny rate will produce £5 640. If the council require £300 000 they will need to charge a general rate of (300 000 ÷ 5640) p = 53·2p.

Income Tax

The government have announced that they will simplify the income tax system in the near future and examination questions will reflect these changes when their exact nature is announced. The examples have been simplified to reduce the amount of calculations involved but illustrate the principles by which income tax is assessed.

A man does not pay tax on the whole of his income. Certain **allowances** are deducted and tax is paid on the remainder. At the moment two ninths of earned income is free of tax. Further allowances include:
1. Personal allowance for the man, and also for his wife if he is married.
2. An allowance for his children.
3. Allowances for dependent relatives.
4. Allowances for Building Society interest, which is payable if the man is buying a house with a mortgage from the Society.

When these and other allowances have been deducted from the man's total income he pays tax on the **remaining taxable income.** The present rate of tax is 38·75%, or 38·75p in the pound.

Example: A man earns £2 700 per year. The first 2/9th of his income is free of tax, and in addition he has allowances as follows:
Personal allowance £600, children's allowance £155,
Building Society interest £261.
He pays tax at 38·75% on the remainder of his income. Find how much tax he pays in a year.

Earned income allowance = 2/9 of £2 700 = £600
Other allowances, £600 + £155 + £261   = £1 016
            Total allowance = £1 616
Net taxable income = £2 700 − £1 616 = £1 084

$$\text{Tax at } 38\!\cdot\!75\% = \pounds\frac{38\!\cdot\!75}{100} \times 1\,084 = \pounds420$$

QUESTIONS

Questions 1–4 refer to the town of Monksworth which has a population of 37 650. The general rate in Monksworth is 45p in the £ and the product of a penny rate is £4 200.

1. The rateable value of Monksworth is:
   (a) £(37 650 × 45)   (b) £(4200 × 45)   (c) £(4200 × 100)
   (d) £(37 650 × 4200)   (e) £(37 650/4200)

2. The total rate collected is:
   (a) £(37 650 × 45)   (b) £(4200 × 45)   (c) £(4200 × 100)
   (d) £(37 650 × 4200)   (e) £(37 650/4200)

3. The rate in the £ necessary for the Drainage Authority requirement of £12 060 is:
   (a) 12 060/45 p   (b) 12 060/4200 p   (c) 37 650/12 060 p
   (d) 12 060 × 45 p   (e) 12 060 × 45/4200

4. The population expressed in standard form is:
   (a) $3 \cdot 765 \times 10^4$ (b) 38 000
   (c) Thirty seven thousand, six hundred and fifty
   (d) $3 \cdot 765 \times 10^{-4}$ (e) $3 \cdot 765^4$

Questions 5–7 refer to the purchase of a refrigerator. The marked price is £48. The terms for Hire Purchase are a deposit equivalent to $16\frac{2}{3}\%$ of the marked price and 18 monthly payments of £2·50.

5. The deposit on H.P. terms amounts to:
   (a) £12   (b) £4   (c) £16
   (d) £8   (e) £6

6. The total paid if H.P. terms are used is:
   (a) £49   (b) £57   (c) £53
   (d) £61   (e) £51

7. If a discount of 12·5% is allowed for cash the cash price is:
   (a) £32   (b) £36   (c) £38
   (d) £40   (e) £42

8. For the normal consumer electricity is charged at 0·88p per unit and a fixed charge of £1·95. If the consumer uses a lot of electricity at night he can have it at a cheaper rate but the fixed charge and day rate are increased—day rate 0·95p, night rate 0·4p and fixed charge £2·73.

A household uses 1 600 units of which one half is used at night time. Calculate their bill on each of the above methods and find which is the cheaper.

9. The following allowances are made before the calculation of income tax: (a) 2/9ths of earned income, (b) personal allowance of £340 for a married man and £220 for a single man, (c) £140 for each child. After allowances the remainder is taxed at the rate of 40p in the £.

Calculate the tax paid by two men, each earning £1 800, one of whom is married with three children, the other of whom is single.

10. The estimated product of a 1p rate in an area is £180,462. What is the total rateable value of the area?

The rate was fixed at 55p, what was the total of the estimated expenses?

The maintenance of public libraries costs a rate of 2·5p. What is the cost of the library service for the year? A householder's property is rated at £150; what is his contribution to the library service?

11. A tape recorder could be bought for £85 for cash or for a deposit of £22·50 and 50 weekly payments of £1·35.

Calculate the difference in cost by the two methods and express this as a percentage of the cash price.

ANSWERS

1(c), 2(b), 3(b) and 4(a). Note that there is no connection between rate calculation and population.

5(d), 6(c) and 7(e).

8. Normal method: £1·95 + 1600 × 0·88p = £16·03
Method allowing cheap night rate:
£2·73 + 800 × 0·95p + 800 × 0·40p = £13·53
Thus using the cheap night rate is cheaper by £2·50.

9. Married man's allowances:

    (a) 2/9ths of £1 800   =   £400
    (b) Married man   =   £340
    (c) Three children   =   £420
                            £1 160

Taxable income £1 800 − £1 160 = £640
Tax payable is 640 × 40 p   = £256

    Single man's allowances:

    (a) 2/9ths of £1,800   =   £400
    (b) Single man   =   £220
                           £620

Taxable income £1 800 − £620 = £1 180
Tax payable is 1180 × 40 p   = £472

10. Each £1 produces 1 p for a 1 p rate.
To produce 18 046 200 p requires £18 046 200. This must be the total rateable value.
If 1 p rate produces £180 462 then a 55 p rate produces
55 × £180 462 = £9 925 410
2·5 p will produce 2·5 × £180 462 = £451 155
A rate of 2·5 p on £150 produces 2·5 × 150 p = £3·75. Thus the householder contributes £3·75 to the library service.

11.   50 weekly payments of £1·35 = £67·50
    Deposit   = £22·50
                       £90·00

The H.P. cost is £5 greater than the cash price.
As a percentage of the cost price this is $\dfrac{5 \times 100}{85}\% = 5·88\%$

Simple Interest

If a sum of money (the **Principal**) is invested at $r\%$ per annum simple interest, the interest gained each year is $\dfrac{r}{100}$ of the principal.

Therefore if the principal is £P the simple interest (£I) in t years is

$$£P \times \frac{r}{100} \times t = £\frac{Prt}{100}.$$

This should be memorised and used as a formula. The inverse statements are $P = \dfrac{100I}{rt}$; $r = \dfrac{100I}{Pt}$; $t = \dfrac{100I}{Pr}$.

The **Amount** is the sum of the principal and interest.

Examples: 1. Find, correct to the nearest penny, the simple interest on £64 for 9 months at 5% per year. Notice that t in the formula is in years. Therefore in this example $t = \frac{3}{4}$.

$$\text{S.I.} = £\frac{64 \times 5 \times \frac{3}{4}}{100} = \frac{64 \times 5 \times 3}{100 \times 4} = £2 \cdot 40$$

2. Find the principal when the simple interest for $2\frac{1}{2}$ years at $3\frac{1}{2}$% p.a. is £27·65.

$$P = \frac{100I}{rt} = 100 \times \frac{553}{20} \times \frac{2}{7} \times \frac{2}{5} = 316$$

∴ the principal = £316

3. Find the rate per cent p.a. when the simple interest on £500 is £30 in $1\frac{1}{2}$ years.

$$r = \frac{100I}{Pt} = \frac{100 \times 30 \times 2}{500 \times 3} = 4$$

∴ Rate = 4% p.a.

The inverse problem of finding P, given the amount A, r and t can be solved by using the formula in the form

$$A = P + \frac{Prt}{100},$$

but it is easier to proceed as in the following example.
What sum of money amounts to £285 at $3\frac{1}{2}$% S.I. in 4 years?
The interest on £100 in 4 years is £14.
∴ £114 is the amount of £100 in 4 years.
∴ £285 is the amount of £100 $\times \dfrac{285}{114}$ = £250

Compound Interest

With simple interest loans the interest is paid periodically, usually every six months. If **the interest is added to the principal** so that the principal increases with time, then the interest is said to be **compound.**

Thus £100 invested for a year at 10% earns £10 interest. If the interest is added to the principal instead of being paid by the borrower the principal at the beginning of the second year is £110. During the second year this £110 earns an interest of:

$$£\frac{110 \times 10 \times 1}{100} = £11$$

If the process is repeated the principal at the beginning of the third year is £110 + £11 = £121 and during the third year the interest is:

$$£\frac{121 \times 10 \times 1}{100} = £12 \cdot 10$$

It can be seen that the interest earned increases each time, and such a system is known as compound interest. In practice the compounding is done more frequently than once a year, often on a monthly basis.

If a large number of compoundings are to be done, then instead of calculating the new principal after each interest period the process may be simplified by using the compound interest formula:

$$A = P\left(1 + \frac{r}{100}\right)^n$$

where A and P are the amount and the principal respectively, r is the rate per cent per annum, and n is the number of years, if the interest is calculated and compounded annually.

Money Conversions

The exact exchange rates between English and foreign money varies from time to time when currencies are revalued or devalued. The rates of exchange for American dollars and French francs are at present:

£1 = 2·54 U.S. dollars   (100 cents = 1 dollar)
£1 = 13·3 French francs   (100 centimes = 1 franc)

Converting from one currency to another requires the use of ratios; the ratio of pounds to dollars is currently 1:2·54 and pounds to francs 1:13·3.

Example: A camera costs 96 U.S. dollars in America and 510 francs in France. Find which is the cheaper to buy, and by how much.

The American camera costs $£96 \times \dfrac{1}{2·54} = £37·82$

The French camera costs $£510 \times \dfrac{1}{13·3} = £38·34$

So the American is cheaper by 52 p.

The Metric System

This is the system of units used in this book. It is used throughout the continent, for all scientific measurements and in many industries in this country. All examination Boards are setting questions on them. The basic unit of length is the **metre (m).** If a larger unit is required then a **kilometre** is used **(km).** Kilo- is the suffix for 1000. For smaller units the **centimetre (cm)**, and the **millimetre (mm)** are used. Centi- means 1/100 and milli- 1/1000.

The basic unit of mass is the **kilogramme (kg).** A larger unit is the **tonne** which is 1000 kilogrammes. For smaller units the **gramme (g),** and the **milligramme (mg),** are used, being 1/1000 and 10^{-6} of a kilogramme, respectively.

The unit of volume should be a cubic metre, but this is rather large for every-day purposes like buying milk, wine and petrol, so the **litre** is used. This is 1000 cubic centimetres.

Shares and Stocks

Shares provide the means both for a person to invest his money and for business concerns to raise capital. Shares are bought and sold through brokers on the Stock Exchange. When shares are first issued they are sold at a certain price, their **nominal value.** As trading continues the price of the shares rises or falls depending on the success of the business which they represent. The owner of the share receives a **dividend,** usually paid twice per year, based on the nominal value of the shares. When this dividend is expressed as a percentage of the market value of the share it is called the **percentage yield.**

For example, to raise £1 000 a firm offers for sale 4 000 shares at 25p each, this being their nominal value. The firm does well and the cost of the shares rise to 75p. At the end of six months the firm announces a dividend of 24%, that is $\frac{24 \times 25}{100}$ p = 6p on each share. The percentage yield on the market price of 75p is thus:

$$\frac{6}{75} \times 100\% = 8\%$$

Stock quoted as $5\frac{1}{2}\%$ stock at 106 means that £100 **stock** (nominal value) would cost £106 **cash** (or $100 stock would cost $106 cash). Again the **income** is based on the **nominal value** – in this quotation $5\frac{1}{2}\%$ – and the yield is the income expressed as a percentage of the cash invested.

Example: How much stock at 110 could be bought for £2 343 and what would be the annual income?

£110 cash buys £100 stock

$\therefore$ £2 343 cash buys £100 $\times \dfrac{2\,343}{110}$ stock = **£2 130 stock**

Annual income = 4% of £2 130

$\qquad\qquad\qquad$ = **£85·20**

QUESTIONS

In questions 1–5 are the statements true or false?

1. The interest on £200 invested at 7% per annum for two years simple interest is £28.

2. The interest on £200 invested at 7% per annum for two years compound interest is £28·98.

3. £1 = 2·54 dollars, £1 = 13·3 francs. Therefore 1 dollar = 13·3/2·54 francs.

4. A share's nominal value is the price it could be bought for on the open market.

5. Stock quoted at 104 means that its cost on the open market has increased by 4% since it was originally offered for sale.

Numbers 6–9 tick the only correct answer.

6. £150 amounts to £186 invested at 6% per annum simple interest in:
   (a) 1 year   (b) 2 years   (c) 3 years
   (d) 4 years   (e) 5 years

7. The compound interest on £5 000 for 2 years at 6% is:
   (a) £5 300   (b) £300   (c) £5 618
   (d) £618   (e) £918

8. The Simple Interest formula is $I = Prt/100$. If this formula is transported to find the time then it becomes:
   (a) $100PIr = t$ (b) $100Ir/P = t$ (c) $100IP/r = t$
   (d) $100/IPr = t$ (e) $100I/Pr = t$

9. Shares of nominal value 75p are priced at £1·05 when a dividend of 10% is announced. A person holding 100 such shares will receive:
   (a) £10·50   (b) £78·75   (c) £7·50
   (d) £10   (e) £82·50

10. Calculate the simple interest on £375 for 1 year and 8 months at 4·5% per annum.

11. After how long does the simple interest on £250 at 5·5% become £22?

12. Find the rate % per annum if the simple interest on £960 for 1 year 3 months is £42.

13. A man borrowed £800 at 6% p.a. simple interest and after 9 months he repaid £500. The rate of interest on the amount still owing was then lowered and after a further 9 months the whole loan was discharged by a payment of £348·60. Calculate the revised rate of interest per annum.

14. Find, to the nearest penny, the compound interest on £780 for 2 years at $3\frac{1}{2}$% p.a., payable yearly.

15. Find, to the nearest penny, the compound interest on £250 in $1\frac{1}{2}$ years at $4\frac{1}{2}$% p.a., payable half yearly.

16. What sum of money invested at $2\frac{2}{3}$% p.a. simple interest amounts to £304 in $2\frac{1}{2}$ years?

17. A man borrowed £650 at $3\frac{3}{4}$% p.a. compound interest and paid back £250 at the end of each of the first two years. How much to the nearest penny did he owe at the end of two years?

18. Find, as a decimal of £1, the difference between the simple and compound interest, payable yearly, for 2 years at $4\frac{1}{2}$% on £1. Hence find the sum on which the difference between the S.I. and C.I. for 2 years at $4\frac{1}{2}$% p.a. is £40·50.

19. A man invested £480 in 25p shares when they stood at 40p and another £480 in the same shares when they had improved to 60p. If the dividend paid was 15% calculate the income from each investment and the percentage yield (to 1 decimal place) on the double investment.

20. The sum of £2 537 was invested in $8\frac{1}{2}$% stock at 118. Find (i) the nominal value of the stock bought, (ii) the annual income from the investment.

21. A man obtained an income of £137·50 by investing £2 900 in $5\frac{1}{2}$% stock. What was the price of the stock?
If he later sold the stock when it stood at 120 and invested the proceeds in $5\frac{1}{4}$% stock at 90, calculate the change in his income.

22. A man sells 1800 £1 shares at £1·30 each and reinvests half the proceeds in 50p shares at 45p which pay a dividend of 4·5% and the remainder in 5·6% stock at 120. Calculate his new income.

ANSWERS

1t, 2t, 3t, 4f and 5t.

Multiple Choice: 6(d), 7(d), 8(e) and 9(c).

10. $£I = £\dfrac{375 \times 1\frac{2}{3} \times 4 \cdot 5}{100} = £28 \cdot 12\frac{1}{2}$

11. $22 = \dfrac{250 \times 5 \cdot 5}{100} \times t$

$t = \dfrac{8}{5}$ Answer 1·6 years

12. $£42 = \dfrac{960 \times 5 \times r}{100 \times 4}$

$r = 3\frac{1}{2}$ Answer: Rate = 3·5% p.a.

13. Interest owing after 9 months $= \dfrac{£800 \times 6 \times 3}{100 \times 4} = £36$

Principal for second 9 months $= £336$
Interest paid for second 9 months $= £348·60 - £336$
$\qquad\qquad\qquad\qquad\qquad\qquad = £12·60$

$$\therefore\ 12\tfrac{3}{5} = \frac{336 \times r \times \tfrac{3}{4}}{100}$$

$\therefore\ r = 5 \qquad \therefore$ Revised rate $= 5\%$ p.a.

14.

$£$	
780	Principal
23·4	$3\%\ (3 \times 1\%)$
3·9	$\tfrac{1}{2}\%\ (\tfrac{1}{2}$ of $1\%)$
807·3	Prin.—2nd yr.
24·219	
4·0365	
835·5555	Amt. after 2 yr.
780	Principal
55·5555	C.I.

Answer $£55·56$

15. $4\tfrac{1}{2}\%$ p.a. $= 2\tfrac{1}{4}\%$ half yearly

$£$	
250·	Principal
5·	2%
·625	$\tfrac{1}{4}\%$
255·625	Prin.—2nd $\tfrac{1}{2}$yr.
5·1125	2%
0·63906	$\tfrac{1}{4}\%$
261·37656	Prin.—3rd $\tfrac{1}{2}$yr.
5·22753	2%
0·65344	$\tfrac{1}{4}\%$
267·25753	Amount
250·	
17·25753	C.I.

Answer $£17·26$

16. Arithmetical method (recommended).
Interest on $£100$ in $2\tfrac{1}{2}$ years $= £2\tfrac{2}{3} \times 2\tfrac{1}{2} = £6\tfrac{2}{3}$
$\therefore\ £106\tfrac{2}{3}$ is the amount of $£100$ in $2\tfrac{1}{2}$ years

$\therefore\ £304$ is the amount of $£100 \times \dfrac{304}{106\tfrac{2}{3}} = £285$

Algebraic method using the Formula $I = \dfrac{Prt}{100}$

Interest $\qquad = £(304 - P)$

$$\therefore 304 - P = \frac{P \times 2\frac{2}{3} \times 2\frac{1}{2}}{100}$$

$$\therefore \qquad P = 285$$

$$\therefore \text{Principal} = \pounds 285$$

17.

£	
650	
19·5	3%
4·875	$\frac{3}{4}$% ($\frac{1}{4}$ of 3%)
674·375	
250	Repayment
424·375	
12·73125	
3·18281	
440·28906	
250	Repayment
190·28906	Balance

Answer £190·29

18.

$$\text{S.I.} = \pounds 0·09$$

$$\text{C.I.} = \pounds 0·092025$$

$$\text{Difference} = \pounds 0·002025$$

$$\text{Sum} = \pounds \frac{40·5}{0·002025}$$

$$= \pounds \frac{40\,500\,000}{2\,025}$$

$$= \pounds 20\,000$$

19. Number of shares in 1st Investment $= \dfrac{480 \times 100}{40} = 1200$

Income from 1st Investment $= \pounds 1\,200 \times \dfrac{25}{100} \times \dfrac{15}{100}$

$$= \pounds 45$$

Number of shares in 2nd Investment $= \dfrac{480 \times 100}{60} = 800$

Income from 2nd Investment $= \pounds 800 \times \dfrac{25}{100} \times \dfrac{15}{100} = \pounds 30$

$$\therefore \text{Total income} = \pounds 75$$

$$\text{Yield} = \frac{75}{960} \times 100\% = 7·8\%$$

20. (i) Nominal value $= \pounds 2{,}537 \times \dfrac{100}{118} = \pounds 2{,}150$

    (ii) Annual income $= \pounds\dfrac{8\frac{1}{2}}{100} \times 2\,150$ using part (i) or

$$= \pounds\dfrac{8\frac{1}{2}}{118} \times 2\,537 \text{ independently of part (i)}$$

$$= \pounds 182{\cdot}75$$

21. $\pounds 137\frac{1}{2}$ was the income from an investment of $\pounds 2\,900$ cash.

$\therefore \pounds 5\frac{1}{2}$ was the income from investing $\pounds\dfrac{2\,900 \times 5\frac{1}{2}}{137\frac{1}{2}} = \pounds 116$

$\therefore$ The price of the stock was 116.

By selling, his capital was increased in the ratio $120 : 116$.

$\therefore$ He invested $\pounds 2\,900 \times \dfrac{120}{116}$ in $5\frac{1}{4}\%$ stock at 90.

$\therefore$ His new income was $\pounds 5\frac{1}{4} \times \dfrac{2\,900 \times 120}{116 \times 90} = \pounds 175$

$\therefore$ Income is increased by $\pounds 37{\cdot}50$.

22. $\qquad\qquad$ Proceeds of sale $= \pounds 1\,800 \times 1{\cdot}3 = \pounds 2\,340$

Number of 50 p shares bought $= \dfrac{2340}{2} \times \dfrac{100}{45} = 2\,600$

$\therefore$ Income from these $= \pounds 2\,600 \times \frac{1}{2} \times \dfrac{4{\cdot}5}{100}$

$$= \pounds 58{\cdot}50$$

Income from stock $= \pounds\dfrac{2\,340}{2} \times \dfrac{100}{120} \times \dfrac{5{\cdot}6}{100}$

$$= \pounds 54{\cdot}60$$

$\therefore$ Total income $= \pounds 113{\cdot}10$

Use of Tables

Tables of squares, square roots, etc., are designed to save time and tedious work in arithmetical calculations, but it must always be remembered that they are **approximate.** If a question asks **for an exact answer tables should not be used.** With four-figure tables the fourth significant figure is not reliable (although the error is small and seldom greater than 1) and **answers** should be corrected to 3 significant figures, but **4 figures must be retained in all working.**

SQUARES

	0	1	2	3	4	5	6	7	8	9	1 2 3	4 5 6	7 8 9
26						7023					16		

The above is an extract from the table of squares. In the left-hand column of the table are the figures 10 to 99. These give the **first two significant figures** of the number to be squared. Then there are 10 columns, known as the **main columns,** headed 0 to 9. These headings give the **third significant figure** of the number; the **fourth significant figure** is to be found at the top of the last 9 columns which are known as the **difference columns.**

To find $26 \cdot 53^2$ we find the digits 26 in the left-hand column and on the on the same row in the main column under 5 we find 7 023. These are the first four significant figures of the square of 265. On the same row and in the difference column under 3 we find 16 and this must be added to 7 023 to give the first four significant figures of the square of 2 653; i.e. we get 7 039.

Now if we evaluated 2653^2 by long multiplication the working would remain the same no matter where the decimal point were inserted in the number. It is therefore necessary now only to determine the position of the decimal point in the square of 26.53. This must be done by a **rough check** (R.C.). Since $26 \cdot 53^2$ lies between 20^2 and 30^2, i.e. between 400 and 900 and the significant figures

57

obtained from the tables are 7093, we can deduce that $26.53^2 = 703.9$ (approx.).

Similarly 0.2653^2 lies between 0.2^2 and 0.3^2, i.e. between 0.04 and 0.09. Therefore $0.2653^2 = 0.07039$ (approx.).

SQUARE ROOTS

$$\sqrt{9} = 3 \qquad \text{but } \sqrt{90} = 9.487$$

$$\sqrt{900} = 30 \qquad \text{but } \sqrt{9000} = 94.87$$

The above examples illustrate that when finding the square roots of numbers with the same digit or digits, **two sets of figures can be obtained depending on the position of the decimal point in the number.** For this reason tables of square roots are arranged either with pairs of rows against each number in the left-hand column, or with pairs of pages. In the latter case the pages are usually headed "square roots of numbers from 1 to 10" and "square roots of numbers from 10 to 100". In finding a square root from tables it is therefore necessary to determine which of the pair to use. This is again done by **inspection**, and if there is any doubt the estimate should be squared to check that it conforms to the original number.

Inspection is carried out by marking off digits **in pairs from the decimal point.**

Thus $\sqrt{9'00'00}$ is 300

and $\sqrt{0.00'00'09}$ is 0.003

Once the first significant figure and the position of the decimal point have been established the correct row or page can then be selected and the square root obtained by the same method as that used for finding the squares.

RECIPROCALS

The reciprocal of a number n is $\dfrac{1}{n}$. As n **increases** its reciprocal **decreases** and for this reason the differences in the difference columns must be **subtracted** instead of added. Apart from this the tables are used in the same way as tables of squares. A **rough check** should always be made to locate the position of the decimal point.

58

LOGARITHMS

The logarithm of a number to base 10 is the index or **the power to which 10 must be raised** to equal the number.

Thus if $10^x = N$ then the logarithm of N to the base 10 is x. This is written more shortly as $\log_{10} N = x$. Any positive base can be used for logarithms, but for arithmetical calculations 10 is convenient. The 10 in $\log_{10} N$ is normally omitted.

Since $10^0 = 1$ and $10^1 = 10$ it follows that **log 1 = 0 and log 10 = 1.** Logarithms of numbers between 1 and 10 increase from 0 to 1 and they have been calculated and set out in the tables of logarithms. Anti-logarithms are also tabulated, and these two tables, both of which are used in the same way as tables of squares are of considerable help in calculations.

To **multiply** two numbers together their logs are **added.**

To **divide** one number by another, the logarithm of the **second is subtracted from the first.**

To **square** a number its logarithm is **doubled,** and to **cube** a number its logarithm is **multipled by three.**

To find the **square root** of a number its logarithm is **divided by two,** and to find the **cube root** of a number its logarithm is **divided by three.**

These rules arise from the laws of indices, mentioned on page 15.

Examples:

$$2 \cdot 432 \times 3 \cdot 178 = 10^{0 \cdot 3860} \times 10^{0 \cdot 5022} \qquad \text{(log. tables)}$$

$$= 10^{0 \cdot 3860 + 0 \cdot 5022} \qquad \text{(indices)}$$

$$= 10^{0 \cdot 8882}$$

$$= 7 \cdot 731 \text{ (approx.)} \qquad \text{(anti-logs)}$$

Answer. 7·73 (to 3 s.f.)

$$9 \cdot 562 \div 2 \cdot 877 = 10^{0 \cdot 9806} \div 10^{0 \cdot 4590}$$

$$= 10^{0 \cdot 9806 - 0 \cdot 4590}$$

$$= 10^{0 \cdot 5206}$$

$$= 3 \cdot 316 \text{ (approx.)}$$

Answer. 3·32 (to 3 s.f.)

Numbers greater than 10

$$24 \cdot 6 = 2 \cdot 46 \times 10$$
$$= 10^{0 \cdot 3909} \times 10^1$$
$$= 10^{1 \cdot 3909}$$
$$\therefore \log 24 \cdot 6 = 1 \cdot 3909$$

In the same way, $\log 246 = 2 \cdot 3909$ and $\log 2\,460 = 3 \cdot 3909$.

This is the main advantage of using the base 10 logarithms. It means that we only need tables of logarithms for numbers between 1 and 10.

The decimal part of a logarithm is called the **mantissa** and the number in front of the decimal point is called the **characteristic.**

Numbers less than 1

$$0 \cdot 246 = 2 \cdot 46 \times 10^{-1}$$
$$= 10^{0 \cdot 3909} \times 10^{-1}$$

If the index were simplified in the normal way it would give the negative number $-0 \cdot 6091$. All logarithms of numbers less than 1 are negative, but for convenience in using tables the **mantissa is always kept positive** and **only the characteristic is made negative when necessary.**

We therefore write $\log 0 \cdot 246$ as $\bar{1} \cdot 3909$. This is read as "bar one point 3 909". The bar is written over the characteristic to show that it is negative while the mantissa remains positive.

In the same way $\log 0 \cdot 0246 = \bar{2} \cdot 3909$ and $\log 0 \cdot 00246 = \bar{3} \cdot 3909$.

Rules for finding and using the CHARACTERISTIC

Count the number of places the decimal point must be moved to put the number in **standard form,** i.e. so that it lies between 1 and 10. If the move is to the right, i.e. the number is less than 1, then the characteristic is negative. If the move is to the left, i.e. the number is positive, then the characteristic is positive.

Anti-logarithm tables are only concerned with the mantissa of the logarithm; the number obtained is in the standard form. The characteristic indicates how many places the decimal point must be moved from the standard form, to the right if the characteristic is positive and the left if it is negative.

Particular care is needed in calculations whenever negative characteristics are present.

Example:
$$\begin{array}{r} \bar{2}{\cdot}1022 \\ +0{\cdot}9814 \\ \hline \bar{1}{\cdot}0836 \end{array}$$

Example:
$$\begin{array}{r} \bar{1}{\cdot}3010 \\ -2{\cdot}4840 \\ \hline \bar{4}{\cdot}8170 \end{array}$$

Multiplying logarithms with negative characteristics again needs care.

Example: $\bar{1}{\cdot}8134 \times 3 = \bar{3} + 2{\cdot}4402 = \bar{1}{\cdot}4402$

Division, when **negative characteristics** are involved, is particularly tricky when the characteristic is not a multiple of the divisor. In this case the characteristic must be made up to the next negative multiple of the divisor by adding to it the necessary negative number, a counter-balancing positive number being added to the mantissa.

Example: $\bar{4}{\cdot}3256 \div 2 = \bar{2}{\cdot}1628$ and presents no problem since 2 divides exactly into $\bar{4}$.

Example: $\bar{2}{\cdot}4316 \div 3 = (\bar{2} + \bar{1} + 1{\cdot}4316) \div 3 = (\bar{3} + 1{\cdot}4316) \div 3$

$\qquad\qquad\qquad = \bar{1}{\cdot}4772$

QUESTIONS

In questions 1–5 are the statements true or false?

1. The product of a number and its reciprocal is 1.

2. The product of two numbers can be found by adding their logarithms and finding the appropriate anti-log.

3. The square root of a number can be found by subtracting its logarithm from log 10 and finding the appropriate anti-log.

4. A logarithm characteristic of 2 is equivalent to 10^2.

5. $\sqrt{3} = 1 \cdot 73$, therefore $\sqrt{30} = 17 \cdot 3$.

Tick the only correct answer:

6. The logarithm of a number is that number expressed:
   (a) in four figures
   (b) correct to four figures
   (c) as a power of ten
   (d) as a power of the number itself.

7. The logarithmic characteristic of a number less than 1 is always a:
   (a) negative number with a negative mantissa
   (b) positive number with a negative mantissa
   (c) negative number with a positive mantissa
   (d) positive number with a positive mantissa.

8. If $\sqrt{450} = 21 \cdot 21$ and $\sqrt{45} = 6 \cdot 708$ then the value of $\sqrt{45\,000}$ is:
   (a) 6·708   (b) 21·21   (c) 212·1
   (d) 2 121   (e) 670·8

9. The reciprocal of 3·7 is:
   (a) 1·370   (b) 0·137   (c) 2·703
   (d) 0·2703   (e) 6·3

10. Given that $\log 4 = 0.6021$ and $\log 9 = 0.9542$ select (a)–(e) to match the logs (i)–(v):

(i)	$\log 36$	(a)	1.2042
(ii)	$\log 3$	(b)	1.5563
(iii)	$\log 16$	(c)	0.3521
(iv)	$\log 2.25$	(d)	0.4771
(v)	$\log 25$	(e)	1.3979

11. Use tables to evaluate:
(i) 724.3^2, (ii) 0.02468^2, (iii) $\sqrt{6.743}$, (iv) $\sqrt{0.8321}$, (v) $10^{0.3417}$, (vi) $10^{2.8196}$, (vii) $1/6.733$, (viii) $1/0.04365$

12. Use the tables to evaluate:
(i) $\sqrt{(23.45^2 + 6.743^2)}$, (ii) $\sqrt{712.3} - \sqrt{230.7}$, (iii) $5.624^2 + \dfrac{1}{0.2156}$,

(iv) $\sqrt{\left(\dfrac{1}{2.413^2} - \dfrac{1}{4.671^2}\right)}$

13. Write down the values of: (i) 4^{-2}, (ii) $9^{\frac{1}{2}}$, (iii) $27^{\frac{2}{3}}$, (iv) 2.1^0, (v) $\sqrt{\frac{1}{64}}$, (vi) $\sqrt{2\frac{7}{9}}$, (vii) $\sqrt{0.09}$

14. Evaluate (i) $\sqrt{[9^{\frac{1}{2}} - (\frac{1}{8})^{-\frac{1}{3}}]}$, (ii) $8^{\frac{2}{3}} \div 4^{-\frac{1}{2}}$

15. If $\dfrac{9.1 \times 10^2 \times 13.2 \times 10^6}{4.29 \times 10^{-2}} = A \times 10^n$, where A is a number in standard form and n is an integer, find the exact values of A and n.

16. Use logs to evaluate: (a) 17.43^3, (b) $\sqrt{\dfrac{19.26}{215.3}}$, (c) $\sqrt[3]{0.2984}$

17. Evaluate, correct to 3 s.f. $\dfrac{3.142\,(6.34^2 - 2.97^2)}{\sqrt{4.2}}$

18. Express $\dfrac{0.007\,162^2 \times 3.854}{76.3^3}$ in the form $A \times 10^n$ where A is a number in standard form, correct to 3 s.f., and n is a negative integer.

19. Calculate the value, correct to 3 s.f. of $\sqrt[3]{\dfrac{0.6253 \times 22.78}{32.43}}$

20. Use logarithms and the compound interest formula, $A = \left(\dfrac{P+r}{100}\right)^n$

to calculate the number of years (to the nearest year), for £650 to amount to £1 000 at $4\frac{1}{2}\%$ p.a. compound interest, the interest being paid yearly.

21. The population of a certain town is 81 000. Assuming that it increases by $3\frac{1}{2}\%$ every year, what will the population be in 10 years time? What was the population 5 years ago?

22. The area Δ of a triangle ABC is given by the formula

$\Delta = \sqrt{s(s-a)(s-b)(s-c)}$, where $2s = a+b+c$.

Find Δ when $AB = 8.4$ cm, $BC = 7.6$ cm, $CA = 12.3$ cm.

In addition to questions set specifically to test knowledge of logarithms there are many questions in which logarithms may be used to simplify and shorten the calculations.

Candidates must notice that logs may not be used in questions,
  (i) where their use is expressly forbidden
 (ii) where an exact answer is required
(iii) where they give insufficient accuracy (four figure logarithms can be relied upon to give answers correct to 3 significant figures only).

ANSWERS

1t, 2t, 3f, 4t and 5f.
Multiple Choice: 6(c), 7(b), 8(c) and 9(d).
Matching: 10. (i) and (b), (ii) and (d), (iii) and (a), (iv) and (c), (v) and (e).
Note that $25 = 100/4$ and $\log 100 = 2.0000$.

11.	(i) $724.3^2 \simeq 700^2 = 490\,000$	Ans. 524600
	(ii) $0.02468^2 \simeq 0.02^2 = 0.0004$	Ans. 0.0006091
	(iii) $\sqrt{6.743} \simeq 2. \ldots$	Ans. 2.597
	(iv) $\sqrt{0.8321} \simeq 0.9$	Ans. 0.9122

64

(v) $10^{0.3417} = 2.197$ 　　　　　(vi) $10^{\overline{2}.8196} = 0.06601$

(vii) $\dfrac{1}{6.733} \simeq \dfrac{1}{7} \simeq 0.1$ 　　　　　Ans. 0.1485

(viii) $\dfrac{1}{0.04365} = \dfrac{100}{4.365} \simeq \dfrac{100}{4} = 25$ 　　　　　Ans. 22.91

12. (i) $\sqrt{(23.45^2 + 6.743^2)} = \sqrt{(549.9 + 45.47)}$

　　　　　$= \sqrt{595.37}$

　　　　　$= 24.4(0)$

(ii) $\sqrt{712.3} - \sqrt{230.7} = 26.69 - 15.19 = 11.5(0)$

(iii) Exp.$^n \simeq 36 + 5$ 　$\therefore$ Exp.$^n = 31.63 + 4.638 = 36.2(68)$

(iv) Exp.$^n = \left(\dfrac{1}{5.823} - \dfrac{1}{21.82}\right)^{\frac{1}{2}}$ 　　$[\simeq \sqrt{(.1 - 0.05)}]$

　　　　　$= \sqrt{(0.1717 - 0.04583)}$

　　　　　$= \sqrt{0.12587} = 0.3549$

13. (i) $\frac{1}{16}$, (ii) 3, (iii) 9, (iv) 1, (v) $\frac{1}{8}$, (vi) $1\frac{1}{3}$, (vii) 0.3

14. (i) $\sqrt{(27-2)} = \sqrt{25} = 5$, (ii) $4 \div \frac{1}{2} = 8$

15. Exp.$^n = \dfrac{9.1 \times 13.2}{4.29} \times 10^{10} = 28 \times 10^{10} = 2.8 \times 10^{11}$

16. (a)

17.43^3	1.2412×3
5291	3.7236

Ans. $17.43^3 = 5290$
(to 3 s.f.)

(b)

19.26	1.2846
215.3	2.3330
2)	$\overline{2}.9516$
0.2990	$\overline{1}.4758$

Ans. $\sqrt{\dfrac{19.26}{215.3}} = 0.299$
(to 3 s.f.)

(c)

$\sqrt[3]{0.2984}$	$\overline{1}.4748 \div 3$
0.6682	$\overline{1}.8249$

Ans. $\sqrt[3]{0.2984} = 0.668$
(to 3 s.f.)

17. (Note. The use of "the difference of 2 squares" enables the whole calculation to be done by logs and this is shorter than by using tables of squares.)

$$\text{Exp.}^n = \frac{3 \cdot 142 \times 3 \cdot 37 \times 9 \cdot 31}{\sqrt{4 \cdot 2}}$$
$$= 48 \cdot 1 \text{ (to 3 s.f.)}$$

3·142	0·4972	
3·37	0·5276	
9·31	0·9689	
		1·9937
$\sqrt{4 \cdot 2}$	0·6232 ÷ 2	0·3116
48·09		1·6821

18. Ans $4 \cdot 45 \times 10^{-10}$ (to 3 s.f.)

19. $\sqrt[3]{\dfrac{0 \cdot 5263 \times 22 \cdot 78}{32 \cdot 43}}$
$= 0 \cdot 718$ (to 3 s.f.)

0·5263	$\overline{1}$·7212
22·78	1·3575
	1·0787
32·43	1·5109
3)	$\overline{1}$·5678
0·7176	$\overline{1}$·8559

20. $1000 = 650 \, (1 \cdot 045)^n$
∴ $\log 1000 = \log 650 + n \log 1 \cdot 045$
∴ $3 = 2 \cdot 8129 + 0 \cdot 0191 n$
∴ $0 \cdot 0191 n = 0 \cdot 1871$
∴ $n = \dfrac{1871}{191}$

Ans. 10 yrs.

1 871	3·2720
191	2·2810
9·795	0·9910

21. (i) $A = 81\,000 \times (1 \cdot 035)^{10}$
$= 114\,000$ (to 3 s.f.)

81 000		4·9085
1·035^{10}	0·0149 × 10	0·149
114 100		5·0575

Ans. In 10 yrs. time the population will be 114 000.

(ii) $81\,000 = P(1.035)^5$

$\therefore P = \dfrac{81\,000}{(1.035)^8}$

$\therefore P = 68\,200$ (to 3 s.f.)

81 000			4·9085
$(1.035)^5$	0.0149×5		0·0745
68 230			4·8340

Ans. 5 yr. ago the population was 68 200.

22. $\Delta = \sqrt{s(s-a)(s-b)(s-c)}$

		Logs
$s = 14.15$		1·1507
$s-a = 6.55$		0·8162
$s-b = 1.85$		0·2672
$s-c = 5.75$		0·7597

$a = 7.6$
$b = 12.3$
$c = 8.4$

$2s = 28.3$ $2s = 28.30$

2)2·9938

31·39   1·4969

$\therefore \Delta = 31.4 \text{ cm}^2$ (to 3 s.f.)

Different Number Bases

We are accustomed to counting in the **decimal system,** which has **ten ciphers,** or numbers – 0, 1, 2, 3, 4, 5, 6, 7, 8 and 9. Any number higher than 9 has to be written in two or more columns.

The first column shows the number of units (10^0), the second column shows the number of 10's (10^1), the third column the number of 100's (10^2) and so on.

The decimal system is most widely used, but its origin is an accident of anatomy, based on our ten fingers. If nature had provided us with only eight fingers no doubt we would only use eight ciphers for counting 0, 1, 2, 3, 4, 5, 6 and 7.

The first column would show the units (8^0), the next column would show the number of 8's (8^1). The next column the number of 8^2 (56 in the decimal system) and so on.

THE BINARY SYSTEM

In recent years the binary system of numbers has become more and more important, particularly when applied to electronic computors. The system only uses the ciphers 0 and 1. A current flowing in an electrical circuit corresponds to the cipher 1, and when there is no current this means 0.

The connection between decimal and binary numbers is shown in the following table.

Decimal Number	Expressed in Powers of 2	Binary Notation
0	—	0
1	2^0	1
2	$2^1 + 0$	10
3	$2^1 + 2^0$	11
4	$2^2 + 0 \ + 0$	100
5	$2^2 + 0 \ + 2^0$	101
6	$2^2 + 2^1 + 0$	110
7	$2^2 + 2^1 + 2^0$	111
8	$2^3 + 0 \ + 0 \ + 0$	1000
9	$2^3 + 0 \ + 0 \ + 2^0$	1001

SIMPLE ARITHMETIC OF THE BINARY SYSTEM

The laws of addition and subtraction can readily be applied to binary arithmetic, remembering that only the numbers 0 and 1 are used. Thus $0+0 = 0$, $1+0 = 1$ and $1+1 = 10$. In this last case the units column is full and we have to carry 1 over to the second column.

Example of addition:
```
      gfedcba
      110110
      110100
     _____
     1101010
```

Columns a and b are straightforward, since $0+0 = 0$ and $1+0 = 1$.
In column c, $1+1 = 10$, the 1 being carried over into column d.
In column e, $1+1 = 10$ as before, the 1 being carried over into column f.
Thus f becomes $1+1+1 = 11$, the 1 being carried over into column g.

Example of subtraction: dcba
$$1011$$
$$\underline{110}$$
$$\underline{101}$$

Columns a and b are straightforward, since $1-0 = 1$ and $1-1 = 0$. In column c we cannot subtract 1 from 0, so must take the 1 from column d, $10-1 = 1$.

The only tables in binary multiplication are that $1 \times 0 = 0$ and $1 \times 1 = 1$. This simplifies the sum as there can be no carrying except in the final addition.

Example:
$$1010$$
$$\underline{\times\ 101}$$
$$1010$$
$$0000$$
$$\underline{1010}$$
$$\underline{110010}$$

Division in binary is again simple, and follows the usual rules. There is no "short division" but "long division" can be carried out rapidly since there is no need for trials – it is immediately clear whether or not the divisor will divide.

Example:
$$\begin{array}{r} 101 \\ 101)\overline{11001} \\ \underline{101} \\ 101 \\ \underline{101} \\ 0 \end{array}$$

OTHER NUMBER BASES

4 days plus 6 days equals 1 week 3 days. Since there are **7** days in the week, additions of this sort can be regarded as examples with base 7 arithmetic. The highest cipher that can exist in base seven is **6,** one fewer than the base.

Similarly 2 ft + 2 ft + 1 ft = 1 yard, 2 ft. In arithmetic with the base 3, only the ciphers 0, 1 and 2 are allowed.

Some examples of addition and subtraction in different number bases are given below:

Base eight: 456
                 +713
                 ————
                 1371

Base six: 425
               +340
               ————
               1205

Base seven: 425
                  223
                 +134
                 ————
                 1115

Base six: 311
               −124
               ————
                143

Obtaining the decimal equivalent from any number base:

Example: What are the following numbers in the decimal system?
(i) 245 with the base six $2 \times 6^2 + 4 \times 6^1 + 5 \times 6^0 =$
                             $= 72 + 24 + 5 = 101$

(ii) 323 with the base four $3 \times 4^2 + 2 \times 4^1 + 3 \times 4^0 =$
                               $= 48 + 8 + 3 = 59$

Mensuration

PERIMETER

The perimeter of any shape is the **total distance round its edges.** In simple cases, such as a triangle, it is calculated by adding together the lengths of the three sides. In some cases, such as that of a rectangle, this procedure may be simplified, since it has two equal lengths and two equal breadths.

$$\text{Perimeter} = 2(\text{length} + \text{breadth})$$

The perimeter of a circle is called its **circumference.** This is difficult to measure accurately but can be calculated if the radius or diameter is known:

Circumference is $2\pi \times$ radius or $\pi \times$ diameter

where π has the value 3·142... or very close to 22/7.

Only part of the circumference may be required. If the circle has centre O and radii OA and OB with an angle θ between them, the problem is to find the arc AB.

The lengths will be in the same ratio as the angles. All the way round a circle is length $2\pi \times$ radius and angle 360°,

$$\therefore \frac{\textbf{Arc AB}}{\textbf{2}\pi \times \textbf{r}} = \frac{\theta}{\textbf{360}°}$$

AREA

The area of **a triangle** can be calculated in two ways:

$$\tfrac{1}{2} \times \textbf{base} \times \textbf{perpendicular height} \quad \text{or} \quad \tfrac{1}{2}\,\textbf{ab sin C}$$

where a and b are the lengths of two sides and C is the angle between them.

The area of **a trapezium is $\tfrac{1}{2}(\textbf{a}+\textbf{b})\textbf{h}$**, where a and b are the two parallel sides and h is the distance between them.

The area of the **circle is $\pi \times (\textbf{radius})^2$.**
Note. This is so often muddled with the circumference formula, but there is no need as areas are always lengths^2.

The area of **the curved surface of a cylinder is $2\pi\textbf{rh}$,** where r is the radius and h is the height of the cylinder.

The area of the **curved surface of a cone is $\pi\textbf{rl}$,** where r is the radius of the base, and l is the slant height of the cone.

The area of the **curved surface of a sphere is $4\pi\textbf{r}^2$,** where r is the radius of the sphere.

The area of a sector of a circle is calculated in a similar way to that of finding the length of an arc, that is by considering the ratios of the areas equal to the ratios of the angles:

$$\frac{\textbf{area of sector}}{\pi\textbf{r}^2} = \frac{\theta}{\textbf{360}°}$$

71

VOLUME

The volume of a **cuboid is length × breadth × height.**

The volume of any **prism is the area of cross section × height.**

The volume of a **cylinder is $\pi r^2 h$**, where r is the radius of the base and h is the height.

The volume of a **sphere is $\frac{4}{3}\pi r^3$**, where r is the radius of the sphere.

The volume of a **pyramid is $\frac{1}{3}$ × area of base × height.**

Note that in all volume calculations there must be **three lengths multiplied together.**

Complex Shapes

Areas and volumes of shapes more complex than the simple ones considered above may be found by "building up" or "breaking down".

Thus the outside area of a cylinder is the area of the curved surface plus the areas of the two end circles:

$$2\pi rh + 2 \times \pi r^2 = 2\pi r(h+r)$$

VOLUME CONVERSION

Problems in volumes often involve **changes in shape,** i.e. a solid metal object being melted and re-cast in another shape, or a liquid being poured from one shaped container to another.

Example: 25 spheres of lead each of radius 5 mm are melted and re-cast to make a disc 4 cm in diameter. Find the thickness of the disc.

$$\text{Volume of lead} = 25 \times \frac{4}{3} \times \pi \times \left(\frac{5}{10}\right)^3 \text{ cm}^3$$

If disc has thickness d cm, volume $= \pi \times \left(\frac{4}{2}\right)^2 \times d \text{ cm}^3$

$$\therefore 25 \times \frac{4}{3} \times \pi \times \left(\frac{5}{10}\right)^3 = \pi \left(\frac{4}{2}\right)^2 \times d$$

$$d = 25 \times \frac{4}{3} \times \left(\frac{1}{2}\right)^3 \times \left(\frac{2}{4}\right)^2 = \frac{25}{3 \times 8}$$

$$= \frac{25}{24}$$

$$\therefore \textbf{Thickness} = 1\frac{1}{24} \text{ cm}$$

Three points to note here:
1. Units must be the same; mm have been changed to cm.
2. The diameter had to be changed to the radius by dividing by 2.
3. Often the calculations can be simplified by cancelling, as here where the two π's cancel.

DENSITY

The density of a substance is **the mass o. a unit volume.**

The formula for calculations is: **Density** $= \dfrac{\textbf{Mass}}{\textbf{Volume}}$

The units are usually kg/m^3 or g/cm^3.
The density of water is 1 g/cm^3.

QUESTIONS

1. Select (a)–(e) to have the same value as 1.–5. The numbers in this question are all BINARY numbers.

1.	$11001 - 1010$	(a)	10011
2.	$101 + 1110$	(b)	11110
3.	110×101	(c)	101
4.	$11001 \div 101$	(d)	1111
5.	$(101)^{10}$	(e)	11001

2. Select (a)–(e) to have the same values as 1.–5.

1.	101000	Base 2	(a)	30 Base 10
2.	1200	Base 3	(b)	35 Base 10
3.	132	Base 4	(c)	40 Base 10
4.	200	Base 5	(d)	45 Base 10
5.	50	Base 7	(e)	50 Base 10

73

3. Select (a)–(e) to match the equations 1.–5.
   1. $3014 - 1032 = 1432$ (a) Base 10
   2. $4450 - 3015 = 1432$ (b) Base 8
   3. $4450 - 3016 = 1432$ (c) Base 7
   4. $2504 - 1032 = 1432$ (d) Base 6
   5. $4450 - 3018 = 1432$ (e) Base 5

4. Select the formula (a)–(e) for the areas of 1.–5.
   1. Rectangle   (a) $\frac{1}{2} \times$ base $\times$ perpendicular height
   2. Trapezium   (b) $\frac{1}{2} \times$ (sum of parallel sides) $\times$ perp. ht.
   3. Triangle   (c) length $\times$ breadth
   4. Parallelogram   (d) $(side)^2$
   5. Square   (e) base $\times$ perpendicular height

5. Select for formula (a)–(e) for the volumes of 1–5.
   1. Cylinder   (a) base area $\times$ height
   2. Cone   (b) $\frac{1}{3} \times$ base area $\times$ height
   3. Pyramid   (c) $\frac{1}{3} \times \pi \times (radius)^2 \times$ height
   4. Cuboid   (d) $\frac{4}{3} \times \pi \times (radius)^3$
   5. Sphere   (e) $\pi \times (radius)^2 \times$ height

6. How many spheres of lead of diameter 2 cm are needed to make a cylinder of radius 5 cm and height 20 cm? The density of lead is $11 \cdot 4$ g/cm^3, what is the mass of the cylinder?

7. Rain falls to a depth of 4 cm on a rectangular field measuring 100 m by 200 m. Find the mass of the water on the ground if 1 m^3 has a mass of 10^3 kg.

8. A pipe has a length 30 cm, it has an external diameter of 5 cm and is 1 cm thick. Find the total surface area of the pipe, inside and out, and the volume of the material needed to make the pipe.

9. A field is in the shape of a regular trapezium whose parallel sides are 160 m and 100 m long and distance 40 m apart. Find the area of the field and the cost of erecting a fence all the way round at £2 per 10 m.

10. A solid metal cone, of base radius 4 cm and height 10 cm, is placed in a cylindrical vessel of diameter 9 cm. Water is poured into

the vessel until it reaches the top of the cone. If the cone is then removed find the depth of the water in the vessel.

ANSWERS

1. 1(d), 2(a), 3(b), 4(c) and 5(e).

2. 1(c), 2(d), 3(a), 4(e) and 5(b).

3. 1(e), 2(c), 3(b), 4(d) and 5(a).

4. 1(c), 2(b), 3(a), 4(e) and 5(d).

5. 1(e), 2(c), 3(b), 4(a) and 5(d).

6. Let N be the number of spheres, then equate the two volumes:
$$N \times 4/3 \times \pi \times 1^3 = \pi \times 5^2 \times 20$$
$$N = 5^2 \times 20 \times 3/4 = 375 \quad \text{(Note that the } \pi\text{'s cancel.)}$$
Volume of cylinder $= 500\pi$ cm^3
Mass of cylinder $= 500\pi \times 11 \cdot 4$ g $= 17\,800$ g

7. Volume of water is $100 \times 200 \times 4/100$ m^3 $= 800$ m^3
Mass of water $= 800 \times 10^3$ kg $= 8 \times 10^5$ kg

8. External radius $5/2$ cm, internal radius $3/2$ cm.
Area of ends $= 2 \times [\pi \times (5/2)^2 - \pi \times (3/2)^2] = 2 \times \pi \times \dfrac{25-9}{4}$
$\qquad\qquad\qquad = 8\pi$ cm^2
Area of outside curve $= 2 \times \pi \times 5/2 \times 30$ cm^2 $= 150\pi$ cm^2
Area of inside curve $\;= 2 \times \pi \times 3/2 \times 30$ cm^2 $= 90\pi$ cm^2
$\qquad\qquad$ Total area $= 248\pi$ cm^2 $= 778$ cm^2

9. Area $= 1/2 \times (160 + 100) \times 40$ m^2 $= 5200$ m^2
Perimeter – the sloping sides each have length 50 m – this is obtained by realising that the two triangles which form the ends have sides in the ratio 3, 4 and 5. Therefore the total perimeter is 360 m and cost $36 \times £2$ to fence – £72.

10. Volume of metal cone $= 1/3 \times \pi \times 4^2 \times 10$ cm^3

Volume of cone and water $= \pi \times (9/2)^2 \times 10$ cm^3

Let the required depth of water be h cm.

Volume of water $= \pi \times (9/2)^2 \times h$ cm^3

$$\pi \times (9/2)^2 \times h = \pi \times (9/2)^2 \times 10 - 1/3 \times \pi \times 4^2 \times 10$$

$$h = 7.3 \text{ cm}$$

Trigonometry

SINES, COSINES and TANGENTS

The following three ratios of an angle θ apply to all right angled triangles. It is usual to letter the angles A, B and C and the sides a, b and c, so that side a is opposite to the angle A and so on.

$$\sin \theta = \frac{\text{opposite side}}{\text{hypotenuse}} = \frac{a}{c}$$

$$\cos \theta = \frac{\text{adjacent side}}{\text{hypotenuse}} = \frac{b}{c}$$

$$\tan \theta = \frac{\text{opposite side}}{\text{adjacent side}} = \frac{a}{b}$$

Notice that $\sin \alpha = \dfrac{b}{c}$, $\cos \alpha = \dfrac{a}{c}$ and $\tan \alpha = \dfrac{b}{a}$

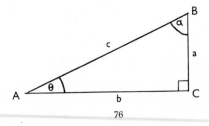

Since $\alpha + \theta = 90$, it follows that:

$$\sin \theta = \cos \alpha = \cos(90 - \theta)$$
$$\cos \theta = \sin \alpha = \sin(90 - \theta)$$
$$\tan \theta = 1/\tan(90 - \theta)$$

USE OF TABLES

Angles are measured in **degrees** and **minutes** (60 minutes = 1 degree). The values of the trigonometrical ratios can be found in tables under the headings of **Natural Sines, Natural Cosines** and **Natural Tangents.** Since the sine and cosine have the hypotenuse of the triangle as their denominator and since the hypotenuse is always the longest side, **sines and cosines are always less than 1.**

In tables the degrees are listed down the extreme left-hand columns of the pages. Other columns are headed in minutes, usually with intervals of 6'. On the extreme right of the page a column is headed **"mean differences"**, these give the corrections to be added or subtracted for 1', 2', 3', 4' and 5'.

NATURAL SINE TABLES

To find the value of sin 24° 33' first find 24° in the left-hand column. Run your finger along this row until you find the value which corresponds to 24°30'—0·4147. Use the "mean difference" column to find the correction needed for the additional 3', that is 8 must be **added** (actually 0·0008).

$$\text{Thus } \sin 24° \, \mathbf{33'} = 0·4155.$$

Using the tables in reverse is a little more difficult.

Sin⁻¹ 0·5894 means the angle whose sine is 0·5894. To find this angle look in the tables to find that number or a number which is just less. In this case the number just less than 0·5894 is 0·5892 corresponding to the angle 36°6'. The remaining 0·0002 is found in the mean difference column on the same row corresponding to 1', so the angle required is 36° 7'.

NATURAL COSINE TABLES

As an angle increases its size from 0° to 90° its sine also increases from 0 to 1. The cosine, however, **decreases from 1 to 0.** This means that when using the cosine tables the **mean difference** value has to be **subtracted**.

To find the cosine of the angle 35° 20′ first find the value for 35° 18′, that is 0·8161. In the difference column the number under 2′ is 3, thus 0·0003 must be **subtracted,** giving 0·8158.

To use cosine tables in reverse look for the number equal or just greater than the one required. To find cos⁻¹ 0·4967, that is the angles whose cosine is 0·4967, we find 0·4970 under the angle 60° 12′. The difference, 0·0003, is found in the difference column of the same row under 1′; thus the angle required is 60° 13′.

NATURAL TANGENT TABLES

As the angle increases from 0 to 90° the tangent increases from **0 to infinity (∞).** Thus as with sines the **mean differences are added.** To save space the integral part of the tangent is only printed in the 0° column, except for very high values when it is changing very rapidly. When the integer changes along a row attention is drawn to it by a different type print. Thus at tan 63° 30′ the integer changes from 1 to 2.

Certain values of these ratios are of special value and worthy of note:

$\sin 0° = 0$	$\cos 0° = 1$	$\tan 0° = 0$
$\sin 30° = 0·5$	$\cos 30° = \sqrt{3}/2$	$\tan 30° = 1/\sqrt{3}$
$\sin 45° = 1/\sqrt{2}$	$\cos 45° = 1/\sqrt{2}$	$\tan 45° = 1$
$\sin 60° = \sqrt{3}/2$	$\cos 60° = 0·5$	$\tan 60° = \sqrt{3}$
$\sin 90° = 1$	$\cos 90° = 0$	$\tan 90° = \infty$

LOGARITHMIC TABLES

The numerical work involved in applications of trigonometry can be considerable, but it is, of course, very much shortened by the use of logarithms. For this purpose we have tables of the logarithms of each ratio and these enable us to obtain the log of a ratio in one reading instead of two. For instance, from the sines table we find that $\sin 56° = 0·8290$, and from the log tables $\log 0·8290 = \overline{1}·9186$, but with a single reading we get from the log sines tables the same result, i.e. $\log \sin 56° = \overline{1}·9186$. These tables can also be used inversely to find the angle whose log sin (etc.) is known. As in the natural tables the differences in the log cosines, log cosecants and log cotangents must be subtracted.

QUESTIONS

In questions 1.–4. pair (a)–(e) with 1.–5.

1. 1. sin 15° 36′   (a) 0·9634
   2. cos 74° 18′   (b) 0·2698
   3. tan 15° 6′   (c) 0·2706
   4. sin 74° 26′   (d) 0·9630
   5. cos 15° 38′   (e) 0·2689

2. 1. sin 21°   (a) cos 69°
   2. cos 44°   (b) cos 21°
   3. tan 70°   (c) cot 20°
   4. cos 46°   (d) sin 44°
   5. sin 69°   (e) sin 46°

3. 1. log sin $\overline{1}$·6500   (a) 24° 4′
   2. log cos $\overline{1}$·9515   (b) 63° 28′
   3. log sin $\overline{1}$·9521   (c) 26° 32′
   4. log cos $\overline{1}$·6500   (d) 26° 34′
   5. log tan $\overline{1}$·6500   (e) 63° 34′

4. 1. sin 60°   (a) $\frac{1}{2}$
   2. tan 30°
   3. cos 60°   (b) $\dfrac{1}{\sqrt{3}}$
   4. sin 45°
   5. tan 45°   (c) 1

                       (d) $\dfrac{\sqrt{3}}{2}$

                       (e) $\dfrac{1}{\sqrt{2}}$

5. In $\triangle ABC$, $AB = 6$ cm, $BC = 9\cdot5$ cm, $\angle B = 47°$ and AL is the perpendicular from A to BC. Calculate AL and $\angle C$.

6. In $\triangle ABC$, $\angle ACB = 133°$, $BC = 6$ cm and the perpendicular, AD, from A to BC produced is 5 cm. CE is the perpendicular from C to AB. Calculate AC, $\angle B$ and CE.

7. R and S are two points on a semi-circle whose diameter PQ is 12 cm, such that PR subtends 52° at the mid point O of PQ, and the chord RS is 5 cm. Calculate PR, $\angle SOQ$ and the length of the arc RS. ($\log \pi = 0.4971$.)

8. A regular decagon (10 sided figure) is inscribed in a circle of radius 12 cm. Calculate (i) the length of a side, (ii) the area of the decagon, (iii) the area bounded by a side and its minor arc of the circle. ($\log \pi = 0.4971$.)

9. $ABCD$ is a quadrilateral in which $\angle A = 90°$, $\angle B = 68°$, $\angle D = 56°$, $AD = 9$ cm, $CD = 5$ cm. CX, CY are the perpendiculars from C to AD, AB respectively. Calculate (i) AX, (ii) $\angle CAD$, (iii) AB.

ANSWERS

1. 1(e), 2(c), 3(b), 4(a) and 5(d).

2. 1(a), 2(e), 3(c), 4(d) and 5(b).

3. 1(c), 2(d), 3(e), 4(b) and 5(a).

4. 1(d), 2(b), 3(a), 4(e) and 5(c).

5. $AL = 6 \sin 47° = 6 \times 0.7314 = 4.3884$
$BL = 6 \cos 47° = 6 \times 0.6820 = 4.092$
$\therefore LC = 9.5 - 4.092 = 5.408$
$\therefore \tan C = \dfrac{AL}{LC} = \dfrac{4.388}{5.408}$

$$
\begin{array}{r}
0.6423 \\
0.7330 \\
\hline
\overline{1}.9093
\end{array}
$$

$\therefore \angle C = 39° 3'$ (or 39° 4')
(Note $\triangle ABC$ is *not* right angled at A.)

6. $\dfrac{5}{AC} = \sin 47° \quad \therefore AC = \dfrac{5}{\sin 47°} = 6.836$
$\therefore AC = 6.84$ cm (to 3 s.f.)
$\quad CD = 5 \tan 47° = 5 \times 0.9325 = 4.6625$
$\therefore BD = 10.6625$

80

$$\therefore \ \angle B = \tan^{-1} \frac{5}{10 \cdot 6625} = 25° 7'$$

$$\therefore \ CE = 6 \sin 25° 7' = 6 \times 0 \cdot 4245 = 2 \cdot 547$$

$$\therefore \ CE = 2 \cdot 55 \text{ (to 3 s.f.)}$$

7. Join PQ. Then $\angle PRQ$ is 1 rt. $\angle$ and $\angle PQR = 26°$.
$\therefore \ PR = 12 \sin 26° = 12 \times 0 \cdot 4384 = 5 \cdot 26(08)$
Join PS. Then $\angle PSQ = 1$ rt. $\angle$ and $\angle SOQ = 2 \angle SPQ$.

$$\text{Sin } \angle SPQ = \frac{5}{12} = 0 \cdot 4167$$

$$\therefore \ \angle SOQ = 2 \angle SPQ = 2 \times 24° 37' = 49° 14'.$$

$$\therefore \ \angle ROS = 180° - (52° + 49° 14') = 78° 46'$$

$$\therefore \ \text{Arc } RS = \frac{78\frac{23}{30}}{360} \times 2\pi \times 6 = \frac{78 \cdot 77\pi}{30}$$

$$= 8 \cdot 249 \text{ (using logs.)}$$

$$= 8.25 \text{ cm (to 3 s.f.)}$$

8. Angle subtended at centre by a side $= 36°$
Length of side $= 2 \times 12 \sin 18° = 7 \cdot 416 \text{ cm}$
Area of decagon $= 10 \times \frac{1}{2} \times 12 \times 12 \sin 36°$
$$= 720 \sin 36° = 423 \cdot 2 \text{ cm}^2$$
Area of segment $=$ area of sector $-$ area of $\triangle$
$$= (\pi \times 12^2 \times \frac{36}{360} - 42 \cdot 32) \text{ cm}^2$$
$$= (14 \cdot 4\pi - 42 \cdot 32) \text{ cm}^2$$
$$= (45 \cdot 24 - 42 \cdot 32) \text{ cm}^2$$
$$= 2 \cdot 92 \text{ cm}^2$$

9. $DX = 5 \cos 56° = 5 \times 0 \cdot 5592 = 2 \cdot 796$
$\therefore \ AX = 9 - 2 \cdot 796 = 6 \cdot 204$
Also $CX = 5 \sin 56° = 5 \times 0 \cdot 8290 = 4 \cdot 145$

$$\therefore \ \angle CAD = \tan^{-1} \frac{CX}{AX} = \tan^{-1} \frac{4 \cdot 145}{6 \cdot 204} = 33° 45'$$

Since $CY = XA = 6 \cdot 204$,
$$YB = 6 \cdot 204 \tan 22° = 2 \cdot 507 \text{ (logs.)}$$
$$\therefore \ AB = 4 \cdot 145 + 2 \cdot 507 = 6 \cdot 652$$

81

BEARINGS

The direction in which one must travel in a straight line to get from
a place P to a place A is known as the bearing of A from P. This
angle can be given in two ways, either as an angle less than $90°$
measured East or West from **the North or South line through P,**
or as an angle measured clockwise from **the North line through P.**
Thus in the diagram the bearing of A from P is N$32°$E or simply
$032°$; that of B is S$20°$E or $160°$; that of C is S$65°$W or $245°$; that
of D is N$45°$W or $315°$.

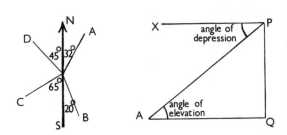

ANGLES OF ELEVATION AND DEPRESSION

These angles are always measured **between the line of sight and
the horizontal.** Thus a man on a cliff at P looking at a boat at A
would measure angle XPA as the angle of depression.

PYTHAGORAS'S THEOREM

Though not strictly trigonometry, Pythagoras's Theorem is included
here since it is relevant to calculations involving right angled triangles.

The theorem states that:

**"The square on the hypotenuse is equal to the sum of the squares
on the other two sides."**

Thus, referring to the right angled triangle on page 76:

$$a^2 + b^2 = c^2$$

SPECIAL CASES OF PYTHAGORAS'S THEOREM

$a^2 + b^2 = c^2$ holds for all right angled triangles, but there are two special cases in which all three sides are whole numbers. These are the 3, 4, 5 triangle and the 5, 12, 13 triangle.

Calculations may often be simplified if you recognise these ratios.

CONSTRUCTING RIGHT ANGLES IN DIAGRAMS

1. Isosceles triangles
If in triangle ABC, $AB = AC$, then the bisector of angle A is perpendicular to BC and bisects BC.

2. Equilateral triangles
The bisectors of all the angles are perpendicular to the opposite sides and bisect them.

3. Rhombus
The diagonals of a rhombus bisect at right angles. They divide the rhombus into four congruent triangles, each of which contains a right angle.

4. Tangents to circles
The radius joining the centre of a circle to the point of contact of the tangent to the circle is perpendicular to that tangent.

5. Chords of circles
The line joining the centre of a circle to the mid-point of a chord is perpendicular to the chord.

6. Semi-circles
The lines joining the extremities of a diameter to any point on the circumference are perpendicular to each other, i.e. the angles in a semi-circle are right-angles.

THE SINE FORMULA

For a triangle in which there is **no right angle** the sine formula can be useful for calculations:

$$\frac{a}{\sin A} = \frac{b}{\sin B} = \frac{c}{\sin C}$$

The sides a, b and c are opposite the angles A, B and C, respectively.

THE COSINE FORMULA

As an alternative to the sine formula it may be found that the cosine formula is a more convenient method for solving a problem:

$$a^2 = b^2 + c^2 - 2bc\cos A$$

Note that this is one of three ways of writing this formula, with b^2 on the left-hand side $\cos B$ will appear on the right and similarly with c.

If the angle is required then: $\cos A = \dfrac{b^2 + c^2 - a^2}{2bc}$

The choice of formula will depend on what information about the triangle you are given, and on what you are asked to find.

OBTUSE ANGLES

For the two formulae given above the angles concerned may be **greater than 90°, that is, obtuse.** Tables of sines, cosines and tangents only involve acute angles, i.e. those between 0° and 90°.
To find the ratios for an obtuse angle, θ, we need to look up the ratios of $(180° - \theta)$.

$$\sin \theta = \sin(180° - \theta)$$

$$\cos \theta = -\cos(180° - \theta)$$

$$\tan \theta = -\tan(180° - \theta)$$

QUESTIONS

1. If the bearing of A from B is $123°$, the bearing of B from A is:
   - (a) $123°$
   - (b) $213°$
   - (c) $303°$
   - (d) $321°$
   - (e) none of these

2. Sin $120°$ has a value equivalent to:
   - (a) $\cos 120°$
   - (b) $-\cos 30°$
   - (c) $\sin 60°$
   - (d) $\cos 60°$
   - (e) $\sin 30°$

3. Cos $100°$ has a value equivalent to:
   - (a) $\sin 80°$
   - (b) $-\cos 10°$
   - (c) $-\sin 80°$
   - (d) $-\cos 80°$
   - (e) $+\sin 10°$

4. The triangle ABC is right-angled at B. BD is an altitude. The length $CD = 8$ cm and $DA = 10$ cm, hence the length CB is equal to:
   - (a) $\sqrt{8}$ cm
   - (b) $\sqrt{10}$ cm
   - (c) $\sqrt{18}$ cm
   - (d) $\sqrt{44}$ cm
   - (e) $\sqrt{80}$ cm

5. The height, to the nearest metre, of a church spire whose angle of elevation is $34°$ when viewed 100 m away is:
   - (a) 55 m
   - (b) 83 m
   - (c) 67 m
   - (d) 148 m
   - (e) 34 m

6. The distance of a buoy, to the nearest metre, from a 50 m cliff, if, from the top of the cliff the angle of depression is $24°\ 18'$, is:
   - (a) 23 m
   - (b) 110 m
   - (c) 108 m
   - (d) 206 m
   - (e) 456 m

7. A man walks 3 km on a bearing N $60°$ E and then 5 km on a bearing N $30°$ W. His distance, to one decimal place, from the start is now:
   - (a) 4 km
   - (b) 8 km
   - (c) $3·4$ km
   - (d) $5·8$ km
   - (e) $1·7$ km

8. The triangle DEF has $DE = 3$ cm, $DF = 5$ cm and the angle $EDF = 30°$. The length of EF is therefore:
   - (a) $2·832$ cm
   - (b) $1·5$ cm
   - (c) 4 cm
   - (d) $2·598$ cm
   - (e) $2·53$ cm

9. The angle of elevation of the top of a vertical flagstaff standing on level ground is found to be $28° 15'$ from a point 7 m above the ground and the angle of depression of the foot of the flagstaff from the same point is $6° 30'$. Calculate the height of the flagstaff.

10. A and B are two posts, 300 m apart, on the same bank of a straight stretch of a river. C is a tree on the edge of the opposite bank such that $\angle BAC = 66° 30'$ and $\angle ABC = 44° 10'$. Calculate AC and the width of the river.

11. A lighthouse L is 25 miles from a port P on a bearing $063°$. A ship leaves P and sails on a bearing $040°$. Calculate its distance from L when it is at A, due west of L. At A the ship changes direction and sails on a bearing $056°$. Calculate the distance AB when it is at B, north-east of L.

12. A launch L leaves a port P and travels on a bearing $072°$ (N $72°$ E) at 12 mile/h. A coastguard station S is situated 3 miles from P on a bearing $200°$. Calculate the distance of L from S 20 minutes after leaving P.

13. In the quadrilateral $ABCD$, $AB = 4$ cm, $BC = 2\cdot3$ cm, $CD = 1\cdot8$ cm, $DA = 2\cdot5$ cm and $\angle A = 56°$. Calculate $\angle C$.

14. A ship left a port A and steamed 46 miles to B at a speed of 12 mile/h on a bearing $127°$ (S $53°$ E). It then changed direction to $243°$ (S $63°$ W) until it reached C, a point due south of A. Calculate the distance BC.
The ship maintained the steady speed of 12 mile/h throughout the whole journey and continued beyond C on the same bearing ($243°$) to D, arriving there 9 h 50 min after leaving A. Calculate its distance from A.

15. ABC is a triangle in which $AB = 8$ cm, $BC = 5$ cm and $AC = 7$ cm. AB is produced to D so that $BD = 4$ cm. DC is joined. Use the cosine rule to obtain an expression for $\cos A$ and hence calculate CD.

ANSWERS

1(c), 2(c), 3(d), 4(d), 5(c), 6(b), 7(d) and 8(a).

9. Let the height of the flagstaff be $(7 + h)$ m and d m the horizontal distance of the observation point from the flagstaff.

Then $\dfrac{d}{7} = \dfrac{1}{\tan 6° 30'}$

and $h = d \tan 28° 15'$

$\therefore \quad h = \dfrac{7}{\tan 6° 30'} \times \tan 28° 15'$

$= 33·01$

$\therefore$ Height of flagstaff $= 40$ m

10. $\angle ACB = 180° - (66° 30' + 43° 10') = 70° 20'$

$\therefore$ In $\triangle ABC$, $\dfrac{AC}{\sin 44° 10'} = \dfrac{300}{\sin 70° 20'}$

$\therefore \quad AC = \dfrac{300 \sin 44° 10'}{\sin 70° 20'} = 222·0$ m

$\therefore$ Width of river $= 222 \sin 66° 30' = 203·5$ m

Answer, $AC = 222$ m; Width $= 203(·5)$ m

11.

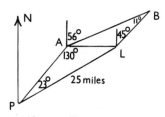

In $\triangle PAL$, $\dfrac{AL}{\sin 23°} = \dfrac{25}{\sin 130°}$

$\therefore \quad AL = \dfrac{25 \sin 23°}{\sin 50°} = 12·75$ miles

In $\angle ALB$, $\dfrac{AB}{\sin 135°} = \dfrac{AL}{\sin 11°}$

$\therefore AB = \dfrac{12·75 \sin 45°}{\sin 11°} = 47·25$ miles

12. $\triangle SPL = 128°$ and $PL = 12 \times \frac{1}{3} = 4$ miles

$\therefore$ By the cosine formula $a^2 = b^2 + c^2 - 2bc \cos A$

$\quad SL^2 = 3^2 + 4^2 - 2 \times 3 \times 4 \cos 128° = 25 + 24 \cos 52° = 39·78$

$\therefore SL = 6·307$ or $6·31$ miles (to 3 s.f.)

13. In $\triangle ABD$, $BD^2 = 2·5^2 + 4^2 - 2 \times 2·5 \times 4 \cos 56°$

$\qquad\qquad\qquad = 6·25 + 16 - 20 \cos 56°$

$\qquad\qquad\qquad = 22·25 - 11·184 = 11·066$

In $\triangle BDC$, $\cos C = \dfrac{1·8^2 + 2·3^2 - 11·07}{2 \times 1·8 \times 2·3}$

$\qquad\qquad\qquad\quad\ 0·4048$
$\qquad\qquad\qquad\quad\ 0·9180$

$\qquad\qquad = \dfrac{3·24 + 5·29 - 11·07}{2 \times 1·8 \times 2·3}$

$\qquad\qquad\qquad\quad\ \overline{1}·4868$

$\qquad\qquad\qquad\quad\ 0·3010$
$\qquad\qquad\qquad\quad\ 0·2553$

$\qquad\qquad = -\dfrac{2·54}{2 \times 1·8 \times 2·3}$

$\qquad\qquad\qquad\quad\ 0·3617$

$\therefore \angle C = 180° - 72° 8' = 107° 52'$

$\qquad\qquad\qquad\quad\ 0·9180$

14.

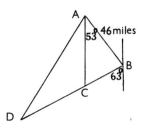

In $\triangle ABC$, $\angle ACB = 63°$

$\therefore \dfrac{BC}{\sin 53} = \dfrac{46}{\sin 63} \qquad BC = 41·23$

Time from A to B $= \frac{46}{12}$ h $= 3\frac{5}{6}$ h

Time from B to D $= (9\frac{5}{6} - 3\frac{5}{6})$ h $= 6$ h $\qquad \therefore BO = 12$ miles

In $\triangle ABD$, $\angle ABC = 64°$

$\therefore AD^2 = 72^2 + 46^2 - 2 \times 72 \times 46 \cos 64°$

$\therefore AD = 66\cdot3$ miles

15. Apply the cosine rule to $\triangle ABC$ to get $\cos A$ and then to $\triangle ADC$ to get $CD = 7\cdot81$ cm.

Note. There is no need to obtain $\angle A$ in this question, as it is $\cos A$ that occurs in both parts.

THREE DIMENSIONAL TRIGONOMETRY

THE ANGLE BETWEEN A LINE AND A PLANE

An angle has two arms and the size of the angle is measured by the amount of turn from one arm to the other. When the angle between a line and a plane is required, it is necessary to identify the two arms. One clearly is the line itself, AP in the diagram. The other arm is a line in the plane. This line is found by "dropping a perpendicular" from any point P on the line to the plane, this cuts the plane at L. The angle required is then $\angle PAL$.

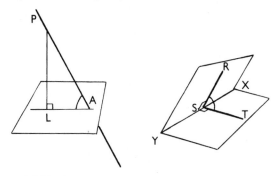

THE ANGLE BETWEEN TWO PLANES

Two planes which are not parallel intersect in a line such as XY in the diagram. From any point S on this intersection draw lines

ST and SR on the two planes at right-angles to XY. The angle between the two planes is the angle RST.

If the plane containing T is horizontal then the line RS is the line of greatest slope on the other plane.

LATITUDE AND LONGITUDE

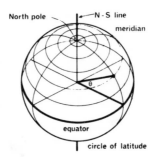

The shape of the earth is roughly spherical and its radius is approximately 6400 km.

Meridians are drawn from the North to the South poles. The **angle of longitude** of a meridian is the angle measured from the centre of the earth around the equator, **the meridian that passes through Greenwich is taken as zero.** Other lines are either East or West of this zero.

Lines of latitudes are small circles parallel to the equator. The angle of latitude is the angle at the centre of the earth made by the equator and the small circle along a meridian.

The equator has thus a latitude of zero and others are either North or South of the equator.

The **radius of a small circle** is the radius of the earth multiplied by the cosine of the angle of latitude.

CIRCULAR MEASURE

Angles are usually measured in degrees and minutes, but in some cases it is easier to use an alternative unit, **the radian.**

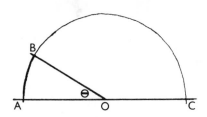

The angle *BOA* in radians is defined by the ratio:

Length of arc AB
Radius of circle

For the semi-circle *ABC* the arc length is $\pi \times$ radius. The angle *ABC* is:

$$\frac{\pi \times \text{radius}}{\text{radius}} \text{ radians} = \pi \text{ radians}$$

Thus we see that π radians is equivalent to $180°$.

Therefore $\quad$ **1 radian** $= \dfrac{180°}{\pi} = $ **57·3°**

Conversely $\quad\quad$ **1°** $= \dfrac{\pi}{180} = $ **0·0174 radians**

Some special cases that are worth remembering:

$360° = 2\pi$ radians $\quad$ $180° = \pi$ radians $\quad$ $90° = \pi/2$ radians
$45° = \pi/4$ radians $\quad\quad$ $60° = \pi/3$ radians $\quad$ $30° = \pi/6$ radians

QUESTIONS

1. $ABCD$ is a desk lid, 100 cm long and with a sloping side of 75 cm. This sloping edge is inclined at $30°$ to the horizontal. The inclination of AC to the horizontal is:

    (a) $30°$ (b) $57°$ (c) $22° 30'$
    (d) $17° 28'$ (e) $16° 42'$

2. Taking the earth's radius as 4000 miles, the radius of the circle of latitude $60°$ N is:

    (a) $\sqrt{2} \times 4000$ miles   (b) 2000 miles   (c) $\sqrt{3} \times 4000$ miles
    (d) $\sqrt{3} \times 2000$ miles   (e) $4000/\sqrt{2}$ miles

3. The distance along the meridian from A; $12°$ N, $10°$ W; to B; $30°$ S, $10°$ W; taking the radius of the earth to be 4000 miles and $\pi = 22/7$, is equal to:

    (a) 1467 miles   (b) 2932 miles   (c) 367 miles
    (d) 5804 miles   (e) None of these

4. $23° 26'$ expressed in radians has a value:

    (a) $0·0067$ (b) $0·4014$ (c) $0·0076$
    (d) $0·4090$ (e) $0·8552$

5. $1·732$ radians expressed in degrees, to the nearest degree has a value:

    (a) $40°$ (b) $57°$ (c) $97°$
    (d) $41°$ (e) $99°$

6. A wall is 6 m long and 2 m high, it runs N.W. to S.E. Calculate the area of the shadow on level ground when the sun is S.W. and at an elevation of $38°$.

7. A and B are two points on the edge of a cliff at the same height above the sea; B is 200 m due North of A. An observer at A notices that a rowing boat at P due East of him is 20 m from the foot of the cliff and its angle of depression is $45°$. Two minutes later an observer at B notices that the boat is now due East of him and its angle of depression is $15°$. Calculate the speed of the boat and the direction in which it is travelling, both assumed constant.

8. A, B and C are three points on level ground such that angle $ABC = 90°$. CD is a vertical pole with D 30 m above C. The angles of elevation of D above A and B are $35°$ and $40°$ respectively. Calculate the distance between A and B.

ANSWERS

1(d), 2(b), 3(a), 4(d), 5(e).

6. The shadow at the end of the wall is at right angles to the wall and is of width $2/\tan 38°$ m $= 2·56$ m

Therefore area of shadow is $2·56 \times 6$ m^2 $= 15·4$ m^2

7. Height of cliff is 20 m, that is AX and BY in the diagram.

$YQ = 20 \times \tan 75°$ m $= 74·6$ m

Therefore $ZQ = 54·6$ m

Therefore $PQ = \sqrt{(200^2 + 54·6^2)} = 207$ m

Therefore the speed of the boat $= 207/120$ m/s $= 1·72$ m/s.

Also $\tan ZPQ = 54·6/200 = 0·273$ Angle $ZPQ = 15° 16'$

The boat is travelling in a direction N $15° 16'$ E.

8. $\angle ADC = 55°$ and $\angle BDC = 50°$

$\therefore AC = 30 \tan 55° = 42·84$

and $BC = 30 \tan 50° = 35·75$

$\therefore AB^2 = AC^2 - BC^2$

(Pythagoras's Th. $- \angle B = 1$ rt.)

$= 42·84^2 - 35·75^2 = 557$

$\therefore AB = 23·6$ m

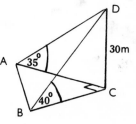

Formulae, Definitions and other Notes

Indices $a^m \times a^n = a^{m+n}$ $a^m \div a^n = a^{m-n}$ $(a^m)^n = a^{m \times n}$

Scales If a linear scale has an R.F. of $1/x$ then the area scale is $1/x^2$ and the volume scale is $1/x^3$.

Direct proportion – two quantities vary but their ratio is unchanged.
Inverse proportion – two quantities vary but their product is unchanged.

Percentage is a fraction with denominator 100.

Profit and loss are calculated with respect to the cost price.

In calculating **Averages** always work with totals.

When drawing **Graphs** always state quantity and units on each axis. Speed is the slope of the line on a distance–time graph.

With **Hire Purchase** the buyer pays a deposit and a number of instalments.

Rates are paid to Local Authorities by property owners. The property has a rateable value. The general rate is the amount that has to be paid for each £ of rateable value.

Simple Interest: $I = \dfrac{Prt}{100}$

Compound Interest: $A = P\left(1 + \dfrac{r}{100}\right)^n$

Metric System: Length   metre (m)

                      Mass   kilogramme (kg)

                      Time   second (s)

To determine a **Square Root** remember to mark off in pairs the digits from the decimal point.

A Logarithm with the base 10 is the power 10 must be raised to give the number, i.e. $10^{0.3010} = 2$ or $\log 2 = 0.3010$

The decimal part of the logarithm is called the **mantissa,** the number in front of the decimal point is the **characteristic.**
The mantissa is always made positive.

Different Number Bases – Decimal system has ten ciphers, 0 to 9. Binary system has two ciphers, 0 and 1.

Note on checking formulae:
Perimeters will have lengths added together.
Areas will have two lengths multiplied together.
Volumes will have three lengths multiplied together.

$$\sin \theta = \frac{\text{opposite side}}{\text{hypotenuse}} \qquad \cos \theta = \frac{\text{adjacent side}}{\text{hypotenuse}}$$

$$\tan \theta = \frac{\text{opposite side}}{\text{adjacent side}} = \frac{\sin \theta}{\cos \theta} \qquad \begin{aligned} \sin \theta &= \sin(180 - \theta) \\ \cos \theta &= -\cos(180 - \theta) \\ \tan \theta &= -\tan(180 - \theta) \end{aligned}$$

$$\frac{a}{\sin A} = \frac{b}{\sin B} = \frac{c}{\sin C} \qquad a^2 = b^2 + c^2 - 2bc \cos A$$

Bearings are measured clockwise from the North line. Angles of elevation and depression are measured from the horizontal.

To measure an angle in **radians** $= \dfrac{\text{arc}}{\text{radius}}$

$$360° = 2\pi \text{ radians}$$

Complete List of
Key Facts Educational Aids

KEY FACTS COURSE COMPANION BOOKS

Physics
Chemistry
Biology
Additional Mathematics
Modern Mathematics
Arithmetic and Trigonometry

Algebra
Geometry
English
Economics
French
Geography

price 55p each.

KEY FACTS CARDS

English Language and
   Examination Essay
English Comprehension and Precis
Geography
History
French
Biology
Chemistry
Physics
Modern Mathematics
Elementary Mathematics
Additional Mathematics
Arithmetic and Trigonometry

General Science
Economics
Geography—Regional
Algebra
Geometry
Technical Drawing
Latin
German
Macbeth
Julius Caesar
New Testament

price 50p each.

KEY FACTS 'A' LEVEL BOOKS

Pure Mathematics
Physics

Biology
Chemistry

price 55p each.

Published in Great Britain by
Intercontinental Book Productions
in conjunction with and available from
SEYMOUR PRESS LIMITED, 334 Brixton Road,
London SW9 7AG.